A UNIFIED FIELD THEORY:

A Human Endeavor

William Foster Day

William2fday@yahoo.com

Printed by CreateSpace

Available on Kindle and other outlets

Available from Amazon.com

First Justification

How can anyone be significant
When we are just one person of
8,000,000,000 people?
How can any individual have a
meaningful life
When at any moment
Our frailty catches us
And stops us dead.
Or we as a species
Are destroyed?

Our individual ideas
Hopes, dreams,
Our personal goals
Crystalize as if from nothing,
Each individual snowflake
Rides the day's currents
Combines with other thoughts
Grows with ……………..

But yet, falling,
Falling

First Justification (cont)

We find our freedom
In these individual, abstract
thoughts
This immaterial abstract domain
With its boundless, infinite
possibilities.

But yet, falling
Falling

Is this moment of feeling
Freedom enough?
The fallen flakes become
A field of whiteness
Maybe covering
A massive void
Of empty space.

Table of Contents

First Justification	3
Preface	7
Description of Aether	10
Quanta Thresholds	12
Dimensions	12
Size	13
Gravitational Constant	14
Dimensions	15
Description Of Bonds	17
Movement	18
Movement Through	19
Doppler Effect	22
Movement In And Because Of The Aether	24
Structured Waves	26
Description of Particles	27
Electrons	30
Tunneling	33
Boson Emissions?	35
Forces	36
Three Proposed Movement Forces	36
$1/r^2$ Forces	37
Weak Force	38
Strong Force	39
Magnetic Force	40
Gravity	43
Gravitational Forces	47
Physical Phenomena Explained By This Theory	53
Michael-Morley Experiments	54
Double Slit Experiments	58
Single Slit Experiments	59
Action at a Distance, Entanglement	61
The Stern-Gerlach Type Experiments	67
Time Dilation Experiments	72

More Detailed Descriptions 81
    Mass 81
    Spin 84
    Electrons 86
    Compton Scattering: 89
    Dislodging of Electrons 89
    Tunneling 91
        Electron Tunneling 91
        Photon Tunneling 93
    Nucleus 94
    Weak Force revisited 96
    Ar > K and Co > Ni 98
    Electricity 100
    Cloud chambers 102
More In Depth Speculations 105
    "Charge!" 105
Fundamental Substance 106
    More Positive Particles Exist 113
    Dark Energy and Dark Matter 116
    Mathematics 118
    Gravitons? 119
Less Definitive Attempts 120
    Sagnac Phenomenon 120
    Moon Laser Experiments 123
    NASA's MMS Experiments 127
    (Lorenz)contraction 129
    A Personal Experiment 132
Conclusion to A Unified Field Theory
    Forward 137
    Why I Created This 139
The Final Justification 146
Appendix 147
    "The 'Hidden Variables' 147
    The Original Rationalization 164
Other Works of William 177

A Unified Field Theory

Preface

It is indeed human to try to understand everything. Not just why we are here; what we are doing in this universe, our place and/or significance, but to try to make sense of it all. How it works. An aspect of this quest is to understand what is this Universe? We know early Greek philosophers (appx. 500 BCE) were speculating. But before them undoubtedly many in the Egyptian, Mesopotamian, and Chinese cultures had philosophies which contained cosmologies and metaphysics. Conceivably a cave person on the African coast with too much time on his or her hands stared at the stars and thought what the hell is going on here.

This is only in part my rationalization and yes … defense … for writing this book. I am fulfilling a human function. But you argue, you are not qualified. Let the experts do it. You are wasting your and our time. (See "Original Rationalization" for a more detailed explanation)

In the early 1970's, during a discussion of the Double Slit experiment in a Philosophy of Science class at Syracuse University, I realized I had a possible answer to the paradox: why a single photon traveling through one or the other slit still creates a wave interference pattern.

Of course, if I had studied Physics, I would have known, that I did not know, and I would have known that all of my explanations for the unexplained physical phenomena were not possible. But I was not educated in this field by our educational system, and so

I did not know. (See Personal notes)

And now over forty years later, I am still adding to my theory. I am still trying to solve the many mysteries and paradoxes posed by present-day physics theories.

On one hand, it is the same answer, a medium through which light travels, and on the other hand it is a deeper explanation and yet in some ways more simplistic.

Even if this model is flawed, as it probably is because of my lack of knowledge and misunderstanding of the results of experiments, concepts, etc., it should point the way to a Unified Field Theory, if such a thing can exist.

My theory involves a medium of massless, polarized units. How can such a thing exist? This will be explored later in the book under the category "More Extreme Speculations."

Description of Aether

What is being proposed is a fabric which is space. Space is not a void but filled by elementary units of massless energy (The Dark Energy?) elastic in quality yet with set parameters of their individual unit's size and strength. The single discrete units have internal vibration in a positive and/or negative dimension, a polarized weaving (Is this even possible? Again, see "More Extreme Speculations"). Therefore, they can attract and repel each other but don't collapse or collide. A positive-negative (bipolar) aether is a third particle, not as large as a combination of the other two. It is a unit because it is a unified configuration and shares energy.

These aethers should not be considered "particles," as such, because they are vibrating pure energy. They are not dots, circles, or spheres, but dimensionless polarization, and only when combined in various ways can create dimension. A single aether's vibration could be caused by the surrounding vibrations of similar and opposite charges. -(Other possibilities are discussed later.) In idealized space the aether-field configuration would be.

Figure 1.

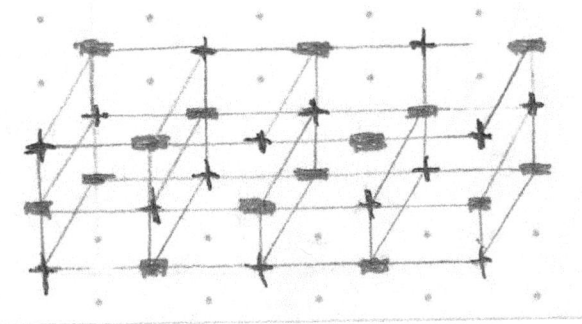

Quanta Thresholds

The fabric of space, the bonds between aethers and the pressure from the aethered universe, create the quanta thresholds. The very fact that action takes place in quanta would suggest that some "thing" prevents a continuous, even flowing event.

Dimensions

All of these polar and bipolar entities are ubiquitous in the Universe. Each of these primary units are a massless single dimension, and only when combining with other units or traveling at the speed of light do they gain dimension. The third dimension emerges when two dimensional objects (a combination of aethers on a x and y axis) are accelerated, or are combined as on a x, y, and z axis. Later in this paper, these configurations will explain why even assuming Super Symmetry more positive particles exists in the Universe than negative.

Size

In one of my earlier papers concerning this theory, Planck's constant (h) 6.626×10^{-34} joules per second was construed as the energy of a single unit of 3-dimensional energy. (To be explored further in section: "More Extreme Speculations.")

The size of a single aether could be approximately 3.99×10^{-35} cm. This is a Planck's Universal unit of least length (Planck) derived from the equation $L_p = (hG/c^3)1/2$. G is the gravitational constant. [To be discussed: next topic. (Notice that the speed of light is cubed.)] Since a single aether is massless, pure energy, then the units will equal the size of vibration. This varies depending upon its surroundings.

An absolute size or time of vibration can not be assigned to individual aethers because in more dense areas, like in and around nuclei of atoms, they are tightly packed, vibrating rapidly in their smallest areas, and in deep space, they expand to their complete capacity.

Gravitational Constant

The Gravitational constant (G_c) is the pressure of the aethered universe. It is the preferred configuration of the Universe.

The Universe (as all systems?) seek an equilibrium. All aethers want to expand but are confined by their surrounding fields and the pressure and configuration of the aethered universe.

Dimensions

Basically, a single polarized unit has no mass, nor does a bipolar unit. Only when combined in a two-dimensional manner and then moving through the medium do they have a demonstrable mass. (Which explains photons as single units.) Of course, three dimensional vibrating configurations have mass. (see further discussion of mass)

In deep, ideal space the attracting and repelling forces are pretty much equalized. (See Fig 1) But it is also the pressure of the aethered universe which is a bond that must be broken in order for action of movement to occur. Whereas around atoms where the attractive forces are great, the bonds of the compressed aether fields need be broken in order for action to occur (a reason why electrons sometimes act like virtual particles: to be explained later). In that each aether unit is indestructible, impenetrable, and variable, indeterminacy is created. The resolution of any measurement is limited by the vibration of individual aethers (Planck's constant). Because the wave and/or particle must "go around" these entities explains why all particles travel in waves and spirals. To add to this indeterminacy, the areas of individual aethers vary in size, rate of vibration, and they can shift on their axis and be moved in space, can change location with other aethers or aether configurations.

Changing one aether will automatically change other aethers. Polarized aethers will attract or repel other polarized aethers and bipolar aethers will shift on their axes.

Description Of Bonds

The bonds among individual aethers are determined by the polar characteristics of each aether plus the collective polarization (the charges) of groups, i.e. aether configurations and structures, particles, and fields. So these individual aethers combine not just to create the fabric of space but all particles.

Particles such as photons, quarks, neutrinos, electrons, protons, etc., have their own structural integrity, plus a unified flow of energy, a configuration of energy and/or vibration. The spinning characteristics of particles in a relatively stationary aether field also help to create the structures' integrities which help to distinguish units from their surrounding aether fields.

They are constructions of individual positive, negative,

and bipolar aethers combined in such a manner to create stability, unity, and in some cases, charge. Because each arrangement vibrates and spins in the medium, it limits the size and possible types and combinations of these particles. The larger particles are connected by the bonds of attraction of individual aethers and by the bonds of groups of aethers i.e. quarks, gluons (to be explained in more detailed later).

Movement

Movement in this medium occurs in two different manners: movement through, and movement in and because of the aether. Particles with mass move through the medium and virtual particles move in and because of it. (A subtle but distinct three-different-movement forces may exist. To be elaborated later)

Movement Through

Since ideal space has innate symmetry, the pushing or pulling of an aether creates compression and vibration waves. An individual aether's space can not be violated, so when an aether is moved, the field is slightly changed. It also can either attract and pull an aether with an opposite charge, or repel and push aethers with similar charges. In an ideal space, symmetrical, the movement would be diagonal in the forward vector. It would push the similar charges which are diagonal to its vector 1) because aether's space can not be violated and 2) the natural arrangement of space is relatively symmetrical with opposite charges surrounding. Moving-charged particles meet little resistance because they attract an oppositely charged field in front of them.

Particles moving through the aether fields can't violate the space of an individual aethers which is equal to Planck's constant and therefore must go around them. This motion creates waves, and this natural configuration of the aether field causes particles to spin and spiral. Not just the physical particle moves through aethers but its charge pulls and pushes the surrounding aethers.

The charge of the particle moves the charged aethers and turns the bipolar aethers on their axis. For example, a projected electron creates a wave both in front of itself and on its sides. A relatively slow moving electron attracts the quickly moving positive and the positive sides of bipolar aethers on all sides of it (to be elaborated "electron's configuration.") The electrical charge projects an aether configuration in front of its trajectory and at the same time physically pulls and pushes aethers to its side. When aethers are moved in this manner, radiation is created. (There should be a slight sign of radiation in front of a projected electron?)

In comparison: pushing a neutral object through the aether would move the aethers in the immediate vicinity but very little to no radiation should be created because aethers are not rotated or even pulled from their locations. Also the length of the waves of neutral projected objects should be shorter. (All aether field configurations in the medium are affected by the $1/r^2$ relationship because of G and the propensity of aether fields to obtain their natural configuration. To be discussed.)

In these cases of charged particles traveling through a field, the field-aethers not only turn on their axis because of the charge but are pulled or pushed because of the mass of the object, as mentioned these movements can create radiation and a Coherently-structured wave. (To be discussed.)

If the projected object is accelerated and moves faster than the aether field in front of it can adapt, then the object seems to gain in mass because aethers accumulate in front of the accelerating object.

Doppler Effect

The aethers in front of an accelerating object become compressed and the light from the object traveling through this compressed aether field becomes a higher frequency. Viewers of this object moving towards them see a "blue shift." And if the object is accelerating away from the viewers, they see a "red shift" because the wake of the aether field is less compressed. Both shifts are correlated to the speed of acceleration.

Movement In And Because Of The Aether

The wave created when particles are projected through the fields and when a photon is created are different. The photon's life is instantaneously created when a wave is created, and a wave is instantaneously created when a photon is created. It is a circular, spinning compression wave (because of the aether field) pulsating outwards from the nucleus.

It may be that when an electron moves to a lower orbit, it creates enough energy to combine positive and negative aethers which results in photons and/or just bipolar aethers are ejected. At the same time the electron creates a compression wave, like an expanding bubble, in which the photon rides. Whether the combining of the aethers is true or not, (to be explored later) energy is released when the electron rearranges the aethers between orbits (see tunneling).

Photons are unified energy configurations. They are a <u>pattern of a particle</u> in the aether field with no rest mass, which means it is a single dimensional aether configuration riding on the wave. Unless this pattern is spinning on the front of the moving wave field, its energy disperses, and it ceases to be a configuration.

Photons are like surfers riding a compression wave. As a wave travels through the ocean, the individual molecules of water (the aethers) do not travel along but just rise and fall (aethers are slightly altered) as the wave passes. And if the wave is dampened, hits a barrier, then the surfer sinks into the ocean, or as in the case of a photon, the energy of the angular momentum ceases.

It would seem to me that if a source of light were moving in a vacuum or void and no medium existed then the speed of the light which originated from this moving source would depend on the speed of the source and would not be a fixed constant; for it would seem that nothing in a vacuum would limit the variations of the speed of light. But the current conception is that the speed of light is a constant, irrespective of the movements of the bodies involved.

In this aether theory the speed of light is only determined by collision of one aether with an adjacent one, so as fast or as slow the light source, the light travels at a constant speed relative to all light sources. Only at the point where one field of aether interacts with another would any variation be detected. (See moon Laser experiments)

Structured Waves

A Coherently-structured wave for a charged particle is

1) a pre-charged wave with a build-up of an opposingly charge in front of the moving particle and

2) the wave expanding perpendicularly with the moving charge and then contracting after its passing.

3) This creates a set spinning wave structure moving in front of the moving object and wave. The waves are in this sense polarized.

Even photons riding in waves create wave and aether field changes (see single slit experiments). This Coherently-structured wave for photons is a recognizable pressure wave in the form of an expanding circle with an internal spinning wave structure.

Description of Particles

As mentioned, each particle, such as a photon, a quark, neutrino, electron, proton, etc., share a structural integrity and a unified flow of energy, a configuration of energy and/or vibration. They are structures connected by the bonds of attracting, individual aethers and by the bonds of groups of aethers i.e. quarks, protons, etc.

All particles are combinations of aether structures and aether configurations. Some are stable structures, and some are temporary configurations. The difference between the two is that aether structures are configurations which naturally united with permanent(?) bonds and the configurations are just patterns in the aether fields (see Cloud Chambers).

Some symbols of protons and neutrons might look like this:

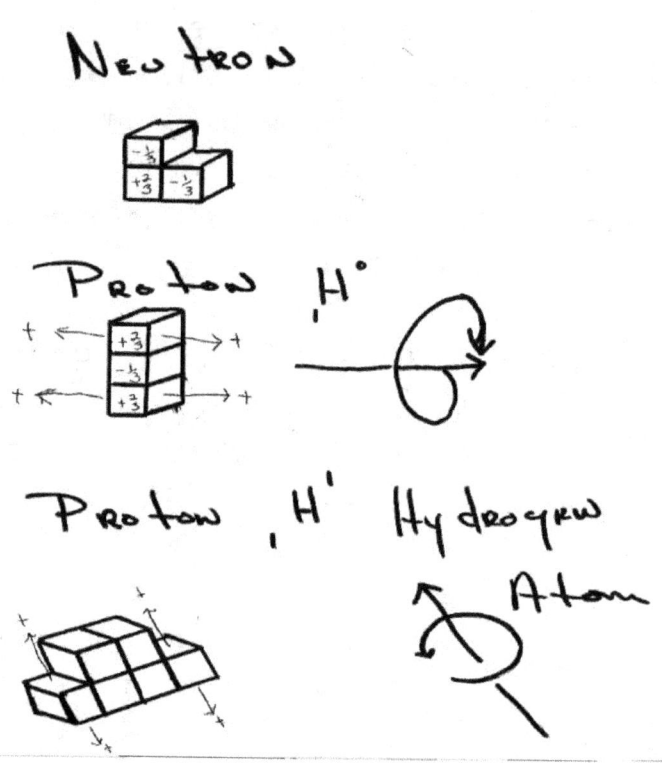

In these simplistic building blocks, notice the symmetry of the proton, which when traveling through the symmetric aether fields would rotate as compared with the neutron which could not freely move through the aether.

An representative form of quarks might be (Fig.3)

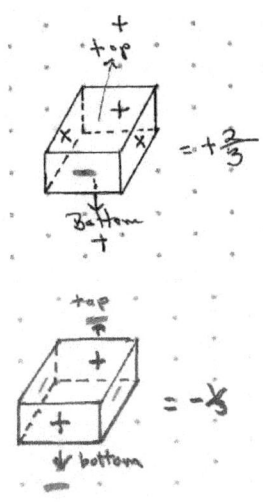

On the bottom, this quark has two faces of the three directional vibrations that are negative, and one that is positive. The above quark has two faces in the positive direction and one in a neutral or bipolar direction. The ironic aspect of these symbols is that the negative one-third quark has twice the mass as the positive two-third quark. It could be that this upper quark only has two positive vibrating directions and not a neutral. And/or the bottom, negative one-third, has twice the number of aethers and/or forces than the positive two-third aether configuration. It may take more aethers to create negative structures.

Electrons

An atom is a vibrating aether structure with aether configurations and structured component parts.

Electrons don't fall into the nuclei of atoms but have discrete orbits (orbits within parameters) because of the condensed, negatively charged aether fields packed against the nuclei. As these fields become less dense, the further away the electron is from the nucleus, the more bipolar regions are created.

The electron obtains orbit at a prescribed distance because the compressed and charged aethers assume a configuration conducive to the structure of the electron. The electron can stay in a set orbit freely because of the size of aethers in the surrounding field, the momentum of the electron, and all the charges reaching an equilibrium. Different patterns of orbits are created because of the arrangement of the polar and bipolar aethers in the fields between orbits and because of the arrangement and movement of the nuclear particles to which the electrons are attracted. The particles in the nucleus can create a polarize arm extending to and holding the electrons in place. These nuclear forces also pull on the electron, encouraging them to change orbits.

Accelerated particles (electrons in orbit) should spontaneously slowdown if the force which accelerates them is eliminated.

But electrons in orbit don't slow down because of their attachment to the vibrating nuclear protons and because their fields are set. The electrons emit no radiation because their surrounding aethers do not change (rotate or change position), either in the direction they travel or to their sides.

As mentioned, the orbital configurations assume a relatively stationary set pattern because the electrons create negative regions of space which again attract and pack positive and positive sides of bipolar aethers as "shells." And yet the orbital space is so conducive that tightly packed and attracting aethers are in front of the orbiting aether structure. So the orbiting "accelerating" electron doesn't disturb the surrounding aether field and cause radiation, yet continually moves in its orbit.

Therefore, with no movement of the surrounding aether field, no radiation is admitted while in orbit. And it takes a quantum force to take it out of the pattern of the aether field.

Maybe one reason electrons and other particles are smears of probability is that it takes many aethers to make them; they are partially virtual, and they move so rapidly. (see "More Detailed Description: Electrons")

Tunneling

Tunneling and creation of photons occurs because electrons changing orbits around the nucleus become a type of virtual particle. The mass of the electron can not penetrate the increasing tapering aether fields between the orbits.

The particle itself doesn't change orbit, but the configuration of the aethers between orbits temporarily conforms to the configuration of the electron. The unified flow of the electron energy passes between orbits. Since it is mass and not just energy like a photon, then the mass of the electron bumps the aethers in front of it, and this continues until an electron emerges in a new orbit. An analogy here maybe pushing a ball into a tube filled with balls.

As this configuration bumps along and spins, it reforms the aether field and creates 1) a displacement-compression wave, radiation, and 2) patterns of/with a unified flow of energy, photons.

(If photons were only contained in the bodies of electrons, a continual emitting of photons would seem to decrease the mass of the electrons, therefore some or all of the photons must originate in the field between orbits.)

The disrupted orbital fields contain mostly negative aethers, especially the fields closer to the nucleus, but both bipolar aethers and positive aethers are in the fields. It is, however, the negative aethers in the field which conflict with the spinning, negative aethers in the electron and create the energy to emit photons.

The dislodged photons ride the compression wave caused by the violent disruption, and the compression wave expands like a bubble. As mentioned, the closer to the nucleus the more compressed and negative the aether field, so undoubtedly the photons nearer the nucleus will take more energy to dislodge and be a higher frequency. (For detailed explanation see Creation of Photons: Compton Effect.)

Boson Emissions?

Question: When accelerated particles emit Bosons, is it always because they are converted to a different particle, or are they displacing aethers and aether configurations? Is there a balanced equation between converted particle plus photons? Or are there more photons than the equations demand?

Forces

Three Proposed Movement Forces

1] When a machine with many cogs and wheels is adjusted, a key turns a wheel and that wheel turns another and so on until the final cog is adjusted. This still takes a finite amount of time to get the Nth wheel to rotate. Relating this to this theory, Aethers structures can change the surrounding aethers: the influence of one aether travels to a second aether then another, etc.

2] Newton's Cradle - A ball hits a row of touching balls: The balls do not move but the pressure is transferred from one ball to the next and the end ball pops up. Sound waves travel at a set velocity through a medium. Notice when one of the internal balls is not touching, the force is dampened. Sound waves do not travel through a vacuum.

3] Instantaneous movement – In a tube filled with balls, when a ball is inserted at one end, a different ball emerges from the other end.

The speed of light is assumed to be two [2].

$1/r^2$ Forces

The causes of inverse-square law could possibly be different for light verses magnetism; therefore, measurements could possibly be subtlety different. With light, the main component seems to be that the surface area of the radiation from the source expands by the square of the radius from that source.

But all electrical, magnetic, and electromagnetic forces are contained within encompassing aether fields and therefore are dampened by those fields relative to the distance from its source. Because of this as a component, they all comply to the $1/r^2$ forces. These forces are the pressure and binding forces of the aether fields and plus G.

Weak Force

All of these orbital pressures help to contain the nucleus of an atom. By combining G_c, the universal constant, and the orbital pressures, these help to determine the weak force. It is the destabilization of the aether fields and the repelling forces in the nucleus which cause beta decay. (For detailed explanation see Creation of Photons: "Compton Effect.")

Strong Force

The protons are aether structures arranged by quarks such that they are positively charged. (See previous figure of quarks). An electrostatic repulsion barrier exists around the protons because they attract negatively charged aethers and the negative sides of bipolar aethers. Gluons (more details later) are strings of aether configurations of negative, positive; negative, positive, etc. then (a) bipolar(s) holding protons to neutrons. The attraction of negative and positive forces in the nucleus could be considered the strong forces. It may be considered that a united energy flow exists between the proton and neutron because of quarks and gluons: patterned aether structures.

An analogy may be the creation of a platoon-bridge so troops and materials can pass between two larger, comparatively stationary bodies of land. The proton would directly attract the negative aethers of a neutron, but the excess positive aethers of both would attract negative aethers and thereby create a bridge. Gluons can be created by bipolar aethers, but since the space is so small they are probably created by individual positive and negative aethers jammed together to create a unity. (See Gluons) This platoon-bridge is a weak analogy because the connection is not to two stationary bodies but two vibrating entities.

Magnetic Force

Flowing electrons (electricity) create magnetic fields because while the stream is continuing, all the surrounding fields become polarized: the bipolar aethers' positive charges face the flowing electrons and of course positive aethers are crammed close to the flow. The polarized aethers create a "highway" for the electrons to travel because of the arranged set structure it creates, with their positive side in front of the flow. As electrons pass, the bipolar aethers not caught in the electron configuration follow their movements and wiggle causing radiations.

An interpretation of the equation C x K☐ = K☐ could be that the power that it takes to set up a magnetic field times the speed of light (the speed of the longitude wave) is equal to the electric constant. So if the flow of electrons is reversed then 1) slight radiation should be created by the initial reversal of all of the polarized aethers and 2) of course, the polarization of the "highway" should be reversed.

Magnetic fields are also constructed by creating an atomic structure which is polarized and can hold an aether configuration. A charged magnetic bar creates and holds a surrounding polarized aether field. The only reason this field is not infinite in all directions is because these aethers in the field attract their opposite charges, and because all aether fields are contained within larger fields. The magnet on my desk is subject to the $1/r^2$ forces.

A moving magnetic bar can cause electrons to flow because within a certain range, the circumscribed field is more cohesive than the inertia of the free electrons caught by the movement of the wave.

Gravity

Physical objects tend to gravitate together, so they can share a more stable and permanent aether field.

In an aethered universe, gravity is defined partially as the strength of an object's aether field which in turn is created by the internal structure of the object, its mass (see discussion of mass). It is the force of attraction, the ability to pull aethers, irrespective of charge, toward it. So mass would partially be defined as the particles of the atoms plus their internal and their external three dimensional aether fields: their set internal structure and their configuration emanating from the internal structure, their aura. The greater the bonds of the particles and aethers, the tighter and more compressed both are and the greater the mass. This may be the reason why, in spite of the number of protons and neutrons, the weights of some atoms are disproportionate to their number of internal particles i.e. Ar> K (see later discussion).

So as mentioned, the closer to the nucleus, the more compressed the aethers. But this doesn't only apply to individual atoms, it applies to larger bodies. All aethers want to expand, fill more space but are compressed by their individual fields and by the encompassing aether patterns including the one created by the Universe. Galaxies have their individual fields as do solar systems and fields held by individual bodies within these systems. Each of these surrounding fields has its own patterns and pressures, but all are contained in the aether field of the Universe with its pattern and pressure.

The encompassing pattern of the larger, stronger fields tends to impose their aether patterns on the aura of objects within their field; therefore, the larger, stronger fields deprive the smaller objects' outer aether fields of their equilibrium. It is the deprivation which is in part gravity. Actually, both the larger and smaller objects want to complete their fields. They want aethers from each other's fields to complete their field. The deprivation is not a deprivation of aethers but a deprivation of the objects to obtain the aether fields they internally generate.

Two fields of patterns will vie for the same aethers and share them, but neither field is complete. These aura fields share a tension. Larger mass objects have stronger aether configurations surrounding them. They use, contort, the smaller object's aether-fields for their own, creating a "need" in the smaller objects aether field, usually more on one side of the object than the other.

This should not be confused with the possible polarity of objects. Since almost all objects have an outer shell of electrons, they attract positive and positive sides of bipolar aethers. In an ideal setting, this outside "shell" is eventually neutralized by positive aethers then negative aethers then positive again creating a layer effect. But as mentioned this aura, the pattern and pressure, eventually dissipates because of the $1/r^2$ forces.

The inverse-square law applies to gravity in two objects' distances from each other because the surface area of their auras expands the further away each object is from the other and because of the Universal Constant G_c,, the pressure and configuration of the Universe's aether field.

Gravitational Forces

Two or more objects create a larger, stronger aether field surrounding all of the objects with a compound privation. The objects together create a larger surrounding field. So, combining objects creates an outer aura which vye for more aethers creating stronger tensions and more deprivation of the fields.

An individual object floating in space will have its field dissipate to the degree that the encompassing aether pattern of the Universe will vye for the external aethers of the object. In that the outer aether field is uniformly deprived on all sides, the object is weightless yet will resist motion.

If the surrounding environment is moving, carrying the floating object as in a galaxy carrying its own aether field, it is relatively motionless, yet moving relative to other objects in the Universe.

(A person in an airplane can be moving relatively to the surrounding environment, outside the airplane, but not moving relatively to the airplane. The person's aether field is not moving so the only direction it is being pulled is downward with the airplane.)

Now when this same weightless object floating in space nears, and comes under the influence of, another object and its gravitational pull, the aura of the first object is still deprived and "wants" to complete its aura potential; so does the encountered object. Both objects tug for the others' aethers: try to use the other's aethers to complete their fields. The number of aethers which are being stressed, pulled, as compared with the number in deep space, is equal to the gravitational force of both. A larger massed object has stronger, set aether structures on all sides and has less of a tendency to be moved.

In other words, the objects trying to complete their internal aether design externally create a tension, elongated, and eschewed aethers. In an attempt to fulfill their internal design, gain their design, the aether field of a larger object "pulls" on the aether field of a second field and causes the smaller object to move. The falling object will have its aura sheared by falling through the larger object's aether field and the gravitational force will increase. The falling object's aether field will resemble a "V" shape as it is sheared.

If this is an object falling to earth, the closer it gets, the tighter the "V" because the aether nearer the earth is more compressed. Undoubtedly the falling objects will become more negatively polarized because the outer layer of aethers will be stripped exposing the relatively negative object: its electron shells. The object will accelerate because the object's field demands more aethers and to some degree nuclear forces, the strong force, come into play. The negatively charged object is attracted to the more positive shells surrounding the earth's electrons surface, the positive shells around the electrons' orbits.

In the fall, a time should come when the negative forces of the electron shells of one object passes the other's electron shell and for a moment repulsion should occur. This could be tested with two extremely finely-honed blocks of metal crashing together. Is there a distance where the objects repel each other? Do they really hit? At what speed?

Also two objects sitting in the same encompassing aether realm, for instance two bowling balls, will both "want" aethers from their surrounding domain and from each other's fields because both are deprived. So even when two similar objects sitting in each other's realm are neutral, or slightly polar, but pretty much the same charge and to the same intensity, seemingly these objects should not attract but should repel. But these two bowling balls sitting in very close proximity will attract each other because each of their auras are incomplete. Each ball vyes for the same aethers in between them and pulls the others' structures. If both of these objects are slightly polar, say negatively charged, then the space between the two objects will repel negative aethers between them and attract positive aethers to the point where the space in between the two is either neutral or slightly positive. This could be measurable.

The deprivation of aethers is on both sides of both bowling

balls. The focus of the deprivation is centered at the core of objects, not only because their internal aether structures set their external form (like a magnetic iron bar forming the surrounding aether structure) but because the object is three dimensional and its aether field is three dimensional. The aethers on both sides of both objects are deprived and are vying with each other. The vector of their force is centered in the object.

A bowling ball resting on the surface of the earth is still pulled "downward" because both the field of the ball and the earth add their tensions.

The deprivation of one aether field with a second is additive. So the aether fields on the external sides of the bowling balls are attracted to each others' centers, and if given the chance, if the force is greater than their attraction to the earth, the two objects will move toward each other. The pressure of the Universe's aether field probably adds to the movement. The pressure between the balls is less than the pressure of the G_c.

When a bucket of water is set spinning either on a rope or just the bucket is spun individually, the gravitational bond between the water and the earth is broken and so not only is the centrifugal force at play to make the water rise up the sides but the gravity of the earth is no longer as active and the gravitational pull of the universe has a larger influence.

Physical Phenomena Which Can Be Explained By This Theory

Michael Morley experiments,

The results of double slit experiments,

Tunneling,

Action at a distance, entanglement,

Gerlach type experiments and the lack of a need for super positional particles,

The discrepancy between periodic numbers and weights of elements,

The airplane atomic clock experiments,

And what we see when we see the results of particle collisions in cloud chambers.

Michael-Morley Experiments

If an aether exists why haven't experiments discovered it, especially the Michelson Morley experiments (Michelson) in the early 1900s, which were specifically designed to find it?

Since the aether is polarized, it is relatively stationary in the earth's magnetic domain, as it would be in all such fields. As the earth rotates, so does its aether sphere. So, the instruments to measure the field and the field rotate at the same time.

As mentioned all objects have their own auras held in place by their gravitational attraction, but magnetic objects have larger domains, more stable, with less slippage when they move. Therefore, as demonstrated, seeking the activity of the aether near the surface of the earth would not produce positive results, unless the testing mechanism passed through the relatively stationary field. (See Sagnac Phenomenon)

Yes, at the turn of the century, Michael-Morley did not discover the aether because it does not slide by any instruments on the earth. The positive, negative and neutral charges are contained in the magnetosphere and rotate as the earth rotates. This can be proven or disproven.

The existence of these individual aether fields like around the sun, earth, and planets should not contradict the statistics which were recorded in the 1919 eclipse experiments and in all of the subsequent like experiments which proved Einstein's theory that light waves are bent when passing through large, massive gravitational fields.

Interestingly enough, the observed data from eclipse were "uncertain to at least 10 percent, and perhaps as much as 20 percent." (Dicke p.27) By re-examining this information one should discover the following: 1) heavenly bodies rotating at different speeds should "deflect" at different variances; 2) the angles of variance on one side of a rotating body should differ from angles on its opposite side. If the light from a single star is deflected with the rotation of the heavenly body, it will have a greater degree of deviation than if the light from the same star if deflected on the opposite side and against the rotation. 3) If two bodies have the same mass but one is magnetic and therefore has a larger and more stable sphere of aether, then this should create a slightly greater variance in the angle of refraction.

And finally 4) it should be discovered that the stronger the gravitational and/or magnetic fields, the greater the compression and therefore the greater the deviation of the angle of a beam of light from a distant star. In the above cases, it is assumed that the longer the duration that the photon exists in a rotating aether sphere, and the more stable and compressed the field, the more it will be diverted.

The exact determinations of differences in variances would be extremely difficult because it would depend on the rotational speed of the object, its aether field, how tightly it is compressed, whether it is magnetized, the slippage of the field, and all of the other disturbances in and around the object. But as in all of my predictions what is to be determined is the "Gross Effect."

Double Slit Experiments

In 1802, when Thomas Young placed two pinholes in a piece of paper, he observed photons (the spinning configurations) plus the wave [compression and radiation] go through both holes, and the fragmented wave interfered with itself. When he closed one hole, the particles and waves went only through one hole: therefore, no interference pattern occurred. In later experiments, when only one photon and wave was released toward the double slits, the wave with a photon went through one slit, and just the wave went through the other, still creating an interference pattern in the aether field. The same wave moving through two different slits skewed it slightly, causing it to interfere with itself.

Single Slit Experiments

In the single slit experiments, when the slit is approximately the size of the wavelength, alternate light and dark bands, Fraunhofer diffraction patterns, are created, with dribble on the sides of the peak:

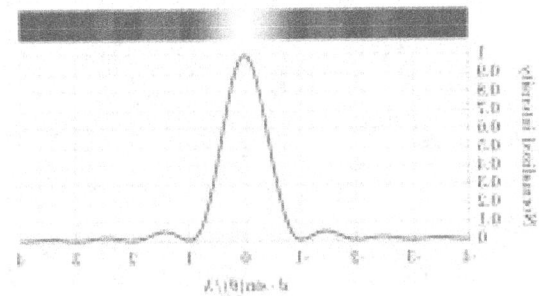

(www.wikipedia.org/wiki/Fraunhofer diffraction).

The wave's energy "will be absorbed and reradiated by the edges of the slit." (Besancon)

This in part creates the dribble because the photon move into a symmetrical field pattern between the slit and the target.

It would seem without an aether medium the photons bouncing off the edges of the slit would create a chaotic pattern? In fact, the photons' paths when altered by any oscillations of atoms (in a crystal) should not create the systematic, uniform, constant pattern it does.

As the size of the slit is increased, the diffraction patterns are decreased because the number of photons which go through the slit increases, thereby creating a more stable flow-pattern in the aether field. Also increasing the intensity of the radiant energy, including using lasers, will change and stabilize the field pattern. This compliance is important to understanding and predicting results from EPR type experiments.

Action at a Distance, Entanglement

When electrons are excited in the Aspect et al type experiments [Aspect, A.; J. Dalibard, and G. Roger, Phys. Rev. Letts 49, 1804 (1982) and Aspect, A.; P. Grangier, and G. Roger, Phys. Rev. Lett. 49, 91 (1982)] twin photons can be created. When the twins are created, they expand outward in their single balloon type compression-radiation wave. Even though these photons travel in a parameter of indeterminacy, within the wave, their polarizations are correlated because the original "tunneling" in the aether field between the orbits creates not only their polarization but the expanding balloon wave in which both travel, and at the same time, this expanding surface of a bubble wave influences the environment in front of the wave. These in themselves are enough to create a correlation which the raw data provide.

In other words, the displacement-compression wave from the electrons changing orbits creates the photons and the compression-radiation bubble wave. In that the bubble is an entity, in itself, the aethers it encompasses as it passes through the field form a set configuration, and this in turn reforms the aether field in front of it.

The actual physical experiments were somewhat different than the EPR *gedankenexperiment*: Not just a single pair was released, but many pairs, a series of "cascades." So as mentioned earlier, just as the flowing photons in a single slit experiment can change the aether field, the flow of paired particles will reform the pattern between the source and the target, thereby increasing the correlations.

But more importantly while all the particles in the expanding balloon are connected in their moment of movement, it is possible to change all of their polarizations by changing the polarization of one particle. If the axis of an aether is shifted by a mechanistic means [traveling through a counter] then the parameter of indeterminacy is reduced, and all axes of all the connected photons on the surface of this bubble would be altered. If this is true, the change in polarization would travel around the bubble at the speed of light or possibly instantaneously (see Three Proposed Movement Forces).

Even if the Aspect (Aspect, A.; P. Grangier) experiment which switches polarized channels at

10ns is truly effective, then it may prove that the "signal" travels faster than the speed of light because the aether in the bubble's surface and the aether between the switches and target have changed instantaneously. This could be an example of pushing a ball into a filled tube.

What I have been implying is that all these types of experiments could be reducing all the photons within a set parameter to a smaller parameter, and therefore the experiment is, in fact, changing the local aether patterns which increases the correlations even more.

Again if this hypothesis is correct, because these photons exist in a parameter of indeterminacy and interference, as in the earlier ocean analogy, both particles will probably be less correlated the further away from each other they travel because of the possibility for interference increases. For instance, as in the double slit experiment, obstacles can disintegrate the unity of the bubble: other waves and particles can interfere. But the bubble or balloon analogy breaks here because once the connection in the compression bubble wave disintegrates, the individual separate parts can still move outward from the source.

A separate experiment could be created to test the above hypothesis. Rather than deploying an off/on switch, a polarized lens could be rotated while the photons are in motion. One, the experiment could start with two lenses with the same polarization, and then one could be rotated. These results could be compared with the number of hits and correlations as the lenses realign. While the slow rotation of the lens is occurring, not only should the correlations progressively decrease, but also the number of hits on the target, for the aether field between the lens and target would be changed. If the lens is stopped when both are realigned, the number of registered hits should increase because the turning filter would have "picked up" photons. In fact, a greater number of hits could be recorded.

It is possible that even the correlations between the hits on the two targets would increase because the field between the source and target is better formed with more particles traveling these paths. But definitely, if this hypothesis is correct, the greater the stream of paired particles, the greater the correlations and the greater the number of hits on the targets.

Summary

Changing the polarization of a coherent wave could be equivalent to pushing a ball into a filled container. All aethers' polarizations move simultaneously along the coherent wave. When one aether's polarization is changed, all the polarization in the whole wave is changed simultaneously.

The Polarized Beam Splitter Experiments

The Stern-Gerlach Type Experiments

This section will refer to Electron's Polarized Beam-splitter Experiments with action through three filters and will explain why after setting the polarization on an x axis in the first filter to eliminate a set spin, then setting the polarization on the y axis in filter #2, that when splitting this flow again in a third filter will give a 50-50% polarization along the x axis again, even though the first filter supposedly eliminated that polarization. This explanation does not need super positional particles to explain it.

To set the scene, in the Stern-Gerlach Experiments atoms were sent through a magnetic field and because of the electrons' charge their paths were diverted to either a plus-one-place or a negative-one-place proving quantization, and that electrons have two different polar spins.

In this aether theory the rearranged electrons travel different paths because of the magnets. As in the double slit experiment, they still travel through the symmetric aether field to their targets, the counter.

In that an electron's spin is quantized is an example of the symmetry of the aether fields. Space is not a continuous event but is broken into units. In earlier experiments, the exploration and discovery were called "space quantization," which seem appropriate to this theory. (Chapter 6)

When both the waves and electrons pass through a beam splitter, half of the polarized electrons and waves go in one direction and the other polarized electrons and waves go in the different direction.

Both patterns should be able to re-emerge at a beam rejoining station before the counter, and a coherent-wave should re-emerge as was: i.e. With the same polarizations.

But when a barrier is placed on one path after the splitter, both the particles and wave pass through the magnetic field which strips one of the orientations and the particles' paths are diverted, and only one of the two polarized spins emerges between the splitter and the rejoinder. One directional spin is eliminated. The coherent wave is disrupted, but the singularly polarized electrons hit the counter.

When these single polarized-spin particles and fragmented wave are passed through a second filter which removes a polarized spin of a different axes, out pops this new coherent polarized wave into a symmetrical aether field that is only somewhat determined by this incoming wave. Again one directional spin was eliminated.

So when these go through a third filter, again the polarized particles make it through, but again only a fragmented wave. And again out pops a new coherent polarized wave into a new symmetrical aether field that is only slightly being determined by this oncoming wave.

Even if one electron at a time is sent through the splitter, after the splitter, fifty percent of those electrons travel on one path and fifty percent in the second direction. If a barrier is placed on the second path, still a polarized electron travels on the first, open, path. But the waves from both still travel in the machinery and interfere with each other. The polarized particle on an x axis from the first path still hits the second filter or counter because this particle helps to create an aether field in front of itself. Therefore, the polarized particle appears.

After the second filter when one half of these series of single electrons are eliminated because the polarization on y axis is stopped by the barrier, it again changes the coherent-wave pattern. The broken waves, waves without an electron traveling the open path, still help form the aether field, especially between the rejoinder and counter screen. This broken wave again must interfere with the electron's and aether fields and thereby randomizes them.

Still the field in front of the re-polarized particle is created by its charge and hits the next filter or counter as polarized.

Because of all of the wave interference, and the symmetrical field after the filter, when this electron travels through a third filter <u>both the configuration of the electron and its Coherently-structured wave</u> are changed and the electron pushing its way through the symmetrical field produces a 50-50 spin outcome on an x axis.

Time Dilation Experiments

Jet Planes with Clocks

Preface

Moving objects have less of an aether field surrounding them because of the shearing action from traveling through larger, relatively stationary, aether field, but when settled, resting (vibrating) in the medium, they regain their field. Once an object, particle, is in motion, it changes the polarity of the aether field in front of it by attracting opposite charges, thereby creates less mass than accelerating particles which press the aether field in front thereby creating greater mass.

Back in the 1970s through the 1990s atomic clocks were placed in jet planes flying east to west and west to east to test the time dilation of the Einstein's theory (Hafele and Keating). A re-examination of data and/or similar experiments should help to prove the existence of a polarized aether surrounding the earth. This theoretical aether would be rotating with the earth because of the earth's magnetic fields. It is assumed that flying against the rotation, from east to west, would indicate a greater slowing of time than flying west to east.

But this does not seem to correspond with the results.

In these experiments, cesium-beam atomic clocks were used and a measurement of the wave length of the exterior electron changing orbits was set as the standard of time.

As mentioned earlier in this aether theory, the pressure created by an object accelerating compresses the aether field in front of the object raising the frequency and thereby would seem to slow time (more than the relativity theories?).

Some Variables

The changes in time because of gravitational fields can be explained: The heavier the gravity fields, the more compressed the aether fields, as if under acceleration; the less intense the gravitation fields the less compressed the aether fields. And in this case of atomic clocks, the more compressed the aether fields around the clocks creates a higher the frequency.

And so in the jet plane experiments, flying at a constant speed, time would not seem to slow down in either direction because the aether fields in front of the atomic clocks would have time to adapt. But flying into the rotating aether field of the earth, the aether fields in front of the clocks would become more compressed and cause the wave length to become a higher frequency. Flying with the rotating aether field of the earth should create less of a compressed aether field around the clocks. (But again this does not seem to be indicated in the results.)

Since the aether around the earth is in a circular structure, the field in front of the jet planes does not prepare the aether field to the degree, the distance, if flying straight into an aether field; less pressure is built. Nevertheless, more pressure would be created flying into the rotating aether and less would be produced by flying with the field.

Variables of time-measurement with Cesium clocks would depend on the speed of the plane, its accelerated take-off and de-accelerated landing, its altitude, and the magnetic fields through which it is flying, but the general description of results should be accurate within parameters and a deviation should exist to predicted results between an eastward and westward flights.

If the planes' altitudes were high enough, where the aether fields were less compressed, flying eastward, into the effect of the flowing (rotating) sun's aether field, might cause time to slow down more than flying with the flow of the sun's radiating aether. If this has any influence on the results then the lower the altitude of the flights, the less the time dilation, because the less influence the sun's aether field would have on the clocks. Also wouldn't seasonal results be different?

The planes flying in either direction, closer to the north and south poles as compared nearer the equator, should experience the greater deviations. Whereas planes flying north-south and south-north should record very little to no variance when compared to each other but should record a greater variance as compared with east and west flights.

Possible Explanations of the Results

In the experiments (Hafele and Keating), the center of the earth was used as the original point of reference. So the comparable clock on the earth is moving west to east, and the planes flying into the aether, east to west would gain time according to relativity. These planes' clocks are not moving as fast as the clock on earth.

And a plane flying with the rotating earth should lose time, since it moves faster than the earth clocks.

Discounting the time dilation due to gravity and the relative speed of the earth-bound clock, a time discrepancy should still exist between the west bound flights, into the rotating aether, and east bound flights, with the rotating aether.

Undoubtedly the planes flying with the rotation of the earth and the fast moving aether should reduce the time greater than predicted with just the movement of the plane because the aether around the clock is even less compressed than the comparable one on earth. But the planes flying into rotating aether should experience a greater slowing of time, more than the other phenomena predicted because the aether in front of the clock is more compressed.

Results published in *Science* 1972

nanoseconds gained

predicted

	gravitational (*general relativity*)	kinematic (*special relativity*)	total	measured
Eastward	+144 ±14	−184 ±18	−40 ±23	−59 ±10
Westward	+179 ±18	+96 ±10	+275 ±21	+273 ±7

Discussion

The eastward bound jets did experience a greater reduction of time than predicted. And even though the measurements of the westward flight seem to be within the parameters of the predictions, it is somewhat slower. The eastward bound airplanes, flying with the aether, have less aether resistance on their clocks and therefore have a lower frequency, and the westward flying planes have a greater than predicted pressure on their clocks creating a higher frequency wave length, and a slower time.

Conclusion

Ever since conceiving the aether, I have overestimated its effect, but it should still have an effect. One of my personal experiments was to rotate a magnet at a high speed for a long time to see if pinholes of light were created in the most sensitive photo film I could find. Of course, I couldn't see any holes, even with a magnifying glass. (See Personal Experiment later in book.)

But still not only does a massless, polarized aether help to explain paradoxical phenomena from physics experiments but when the above experiments have deviations from the theoretical predicted results, these results may be explained by this aether, not only with the time dilation experiments but with the moon laser experiments as well. (See Moon Laser Experiment discussion later in book.)

More Detailed Descriptions

The following will be a more detailed description of Mass, Spin, Electrons, Compton Scattering, Tunneling, Atoms, and phenomena in Cloud Chambers.

Mass

Mass could be defined as the ability to move aethers and change an aether field.

In that particles are unified energy structures, vibrating and moving in a relatively stationary field, they display mass. As an example: neutrons do not have charge, but yet they have mass. Their mass is their internal structure of aether configurations: quarks, and to some degree, the energy which these exert on each other (gluons) and the surrounding aura which they establish. The greater the bonds of the particles and aethers, the tighter and more compressed both are and the greater the mass. This may be part of the reason why, despite the number of protons and neutrons, the weights of some atoms are disproportionate to their number of internal particles. (See section Ar> K and Co>Ni.) It would seem that the higher the frequency of the atom, proportionately lower the mass of that atom, because the more rapidly vibrating atoms would undoubtedly shed more of their surrounding aether structures. Do Argon and Cobalt vibrate faster than Potassium and Nickle?

The aura, no matter how small, extends into the surrounding aether field and creates inertial mass. Particles do not move without an external force. Moving-objects have less aether fields surrounding them, but when settled regain their fields. Once an object, particle, is in motion, it changes the polarity of the aether field in front of it, thereby creates less mass resistance. Accelerating particles press the aether field in front. The aethers in front of the accelerating particles don't have time to respond thereby create greater mass because of their aether build-up in front of the objects. This also creates The Doppler Effect.

Spin

Some particles spin, individually. Some of these create larger and larger units, which include their surrounding aether fields and some of these areas spin. In fact, all matter, particles, could be spinning in a relatively stationary aether field which helps to create a force that holds the units together. The spinning separates the unit from the field.

Electrons spin relatively to and because of their environment and have an internal spin.

If a charged particle (like a free electron) travels through the medium of space, it displaces aethers. Attracted by the positive and repelled by the negative aethers in the field through which it travels, it spins. The spin exists because all moving configurations and particles have to go around aethers and in some cases, configurations.

An electron's spin could be considered 1/2 because its movement within an equivalent space of an aether field is only one-half a total spin. It takes two equivalent areas of space to return the electron to its initial position.

The two different directions of electron's internal spin arrange its surrounding aethers' environment creating magnetic polarized fields surrounding itself. A magnetic moment is produced. Two different electrons can create opposing aether fields that rotate with the electrons but can not violate each other's space.

The internal spin may be caused by its vibration in an aether field, and the vibrations may be caused by the interaction between positive and negative aethers.

Electrons

One of the main problems with this theory is the counter effect on particles moving through aether fields. For example, when an electron plows its way through an aether field, it will attract positive aethers on its side and leave a wake of positive aethers pointing at the passed electron. Why don't these aethers slow, if not stop, the electron. Still as mentioned, the moving electron transposes the field in front of it into its opposite configuration by attracting positive aethers and turning bipolar aethers on their axes then compresses this future field.

Also, as explained, in an orbit around the nucleus of an atom, layers of packed aethers create the "shells."

But with freely moving electrons, a different explanation is needed. The structure of the electron itself may answer this problem.

Electrons' configuration could look like this, spinning through an aether field:

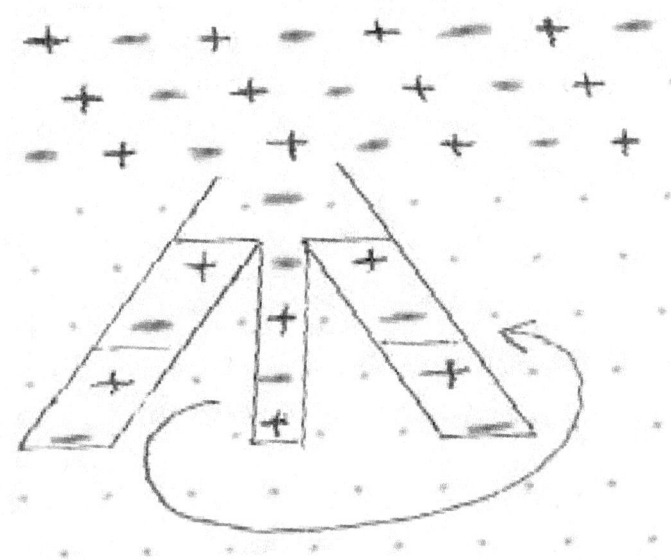

With a combination of bipolar and positive aethers holding the negative aethers in place on the circumference of the electron particle, the natural structure of the particle and the field causes the electron to spin.

The two-dimensional electron spins forward becoming three dimensional. It is partially a virtual particle in that when it spins its configuration on its circumference-wings pass through the aether field. (see section: tunneling). It does not change the field. Also a bipolar aether on the end of the particle causes it not to attract positive aethers.

When the electron's aether structure is not compliant with the surrounding aether field it dislodges radiation. Both in free flight and when changing orbits. When changing orbits around the nucleus where the fields become more and more compressed, and the electron becomes less compatible with the fields.

It may be that violent consequence of the orbit intrusion is not just to dislodge a bipolar aether, a photon, but to combine a positive and a negative aether to create a photon.

So with this configuration and motion, the three dimensional electron only exists when moving (accelerating?) pushing forward (a) negatively polarized aether(s)? Does this explain a vibrating electron?

Compton Scattering:

Dislodging of Electrons

Electrons can be freed from atoms in two different manners: A direct hit by a photon or other particles and a disruption of the aether field in which the electron is held.

It may not only be the increased energy given to the electron (by the hit of a photon) but the destabilization of the aether field between the orbits. Examples: charged particles flying by close enough to disrupt the aether fields and yet not hitting the electron. Yes, a loosely held electron can break its confines by particles crashing into its surrounding aether field or a charged particle pulling the surrounding aether field apart. In both cases the vibrating electron can be set free.

In dislodging electrons in orbit, the greater the frequency of the penetrating light, the closer to the nucleus the light can penetrate through the packed aether field, and therefore the greater the frequency of the dislodged electron which has to pass back through this aether field. So, the closer to the nucleus, the dislodged electron is, probably the greater the repelling force because the concentration of negative aethers which surround the nucleus would repel electron at a greater speed than an electron from an outer orbit. (????) This is a question because of the attraction of the proton could neutralize this force.

Tunneling

Electron Tunneling

Electrons changing orbits around the nucleus become a type of virtual particle. The particle itself doesn't change orbit, but the configuration of the aethers between orbits temporarily is forced to conform to the configuration of the electron, and only the unified flow of electron-energy pushes between orbits. Since it is mass and not just energy like a photon, then the mass of the electron spins into the aethers, and this continues until an electron emerges in a new orbit. An analogy here may be filling the room with hardened racquet balls. Pop one in at the door and a different one will emerge from a window.

As this configuration bumps along and spins, it reforms the aether field and creates 1) a displacement-compression wave, radiation, and 2) patterns of/with a unified flow of energy, photons.

The spinning massed-electrons "kick out" bipolar (and probably positive) aethers from the aethered configuration between the orbits to complete the orbital shift. The photons, the bipolar aethers, ride the compression wave caused by the violent disruption of the field and the compression wave expands like a bubble.

I can believe that when an electron transforms to a different particle it can "give-off" photon(s), but if the same electron "gives-off" photons over and over by changing orbital levels, where does it get its new energy to do so? Why doesn't it change its mass?

No, the photons are mainly created from the aether fields between orbits.

Photon Tunneling

The wave and the configuration of a single photon must travel in an aether field which is conducive to both. But when a photon tunnels, the photon configuration finds a path it can travel through a barrier. Undoubtedly the wave is stopped, and the photon will be reshaped because the aether field between the atoms is in a constant state of motion. Since the protective wave is destroyed, the shape of the photon packet is dependent on the configuration of the medium it meets and its vibrations.

After passing through the barrier, the "tunneled" photon's wave packet is smaller and "'reshaped.'" (Chiao)

Rather than a photon having its full amplitude, it may be compressed and emerge with a different amplitude. Also do the emerging photons have varying energies? By changing the structure and/or elements of the barrier, the parameters of the energy and amplitude should change, because the emerging photons would be dependent on the varying vibrating aethers within the material.

Nucleus

How can protons and neutrons create a unified energy structure, the nucleus? The nucleus is very compact. So the aethers connecting the protons and neutrons create a strong set pattern. How is this possible?

The structural configuration of the quark is met with its opposite aether configuration, a gluon, and this gluon configuration is repeated and repeated until at the end of the "string" is its anti-quark, a different colored quark.

Some gluons holding quarks together are the same aether configurations repeated over and over again. And some could be bipolar aethers with extra negative aethers attached, arranged from the proton as negative-bipolar, negative-positive, etc. to the negative dimensional vibration of the quarks of the neutrons. They gain an attractive force because of the extra disproportional opposite aethers attached.

These aether configurations only predominately exist in the nuclei where the two particles have created them.

These forces do not diminish with distance but gain as compared with a dampening force (the Weak Force) which is created when a particle just attracts its opposite charges and the opposite sides of bipolar aethers.

Weak Force revisited

As mentioned, the weak force is mainly created by the pressure of the aether fields. Nuclei can contain an inordinate number of protons, not only because of the strong force and gluons, but because of the pressure of the surrounding fields and the aethered universe. The compressed fields surrounding the atoms need to "give way" for radioactive decay to occur.

Naturally occurring radio-activity can be both the transmutation of atoms, their internal instability, and the decay of their surrounding field. If the surrounding aether field has a disproportionate number of charged and/or bipolar aethers, it encourages decay. Or if the field is disturbed, ripped, by some outside force i.e. a positive or negative passing particle, then decay of the nucleus takes less energy and is more probable.

Alpha particles can break through because of the nuclear volatility and/or their unstable surrounding field. They either pass all the way through the predominately negative shell or disintegrate like dead flesh being eaten by parasites (the negative surrounding aether field helps neutralize their charge).

This phenomenon occurs in a very short distance from the nucleus, in the surrounding aether field. But it can destabilize the surrounding field even further. The remaining particles in the nucleus readjust: The aethers holding the particles together and the particles themselves shift on their axes and/or expand or contract. In the same way electrons changing orbits rearrange the intervening aether fields, these reshufflings cause gamma radiation.

Ar > K and Co > Ni

The masses of some atoms are disproportionate to their atomic number: their number of internal particles, i.e. Ar > K and Co > Ni. In both cases, in spite of gaining a proton, their atomic masses decrease. The most extreme example is from Argon to Potassium. Potassium gains a proton and an electron. And even though potassium's isotopes have more neutrons, the atomic mass of potassium is less than Argon. Also just in the physical aspect of this phenomenon, generally Argon is a gas and potassium is a liquid. So, why and how can potassium have less mass?

The obvious answer according to this theory is both Argon and Cobalt contain more gluons, more aethers, more energy than Potassium and Nickel. Their internal configurations, the arrangement of their protons and neutrons, are more conducive to holding more, and less aethers, and more and less energy. Also their internal configurations may create larger and stronger auras in their surrounding fields. To compliment these structures, Argon is even a larger atom than Potassium, and therefore should be able to contain more aethers. This may explain why from Cobalt to Nickel, not as much a discrepancy in atomic weights exists as from Argon to Potassium. The sizes and internal structures of Cobalt and Nickel are more identical; their isotopes have a difference of one neutron, and both are metals. To repeat an earlier question: It would seem that the higher the frequency of the atom, proportionately the lower the mass of that atom. Do Argon and Cobalt vibrate faster than Potassium and Nickel?

Electricity

Flowing electrons create polarized fields because they attract the positive sides of bipolar and positive aethers, and these fields remain stationary as long as the flow maintains. The polarized aethers which create the "highway" for the electrons to travel become arranged in a set structure with their positive sides facing the flow. The negative moving force attracts a positive force in front of its flow. As electrons pass, the bipolar aethers follow the moment and the wiggling causes radiation.

An interpretation of the equation C x K☐ = K☐ could be that the power that it takes to set up a magnetic aether field times the speed of light (the speed of the longitude wave) is equal to the electric constant.

So if the flow of electrons is reversed then 1) the polarization of the "highway" should be reversed and 2) more radiation should be created by the initial reversal.

When the electricity is shut off, the natural configuration of the surrounding aether field will predominate, and the polarized field will cease. Another prediction: when the current of DC is reversed, radiation should occur. If the flow is reversed, not only will the polarization be reversed but radiation should be emitted to the degree that these bipolar aethers need to move, rotate.

With an alternating current (AC), an aether field would be created, but only a lesser amount of radiation should be emitted after a couple of cycles. The aether field further from the flow should be set and so more radiation should be closer to the flow and an smaller amount of radiation should be further from it (than DC current). This slight amount of radiation should occur due to the rotation of the bipolar aethers on their axes each time the current reverses its direction. But a more permanent field should be set further away. So a continual flow of DC should cause less radiation than AC, switching direction of DC should cause greater radiation than a single cycle of AC, and the first couple of cycles of AC should create a greater amount of radiation than repeated cycles.

Cloud chambers

"Nothing comes into being out of what is non-existent."

Epicurus

By accelerating or colliding particles in the space of a cloud chamber field, any configuration can be dislodged, whether it would exist naturally as a unified particle or not.

Preparing this space and/or colliding two distinct particles would change the probability of chance-occurrences of "seeing" certain aether configurations. But what is being seen is any combination of positive, negative, and bipolar aethers structures that can exist, including "temporary particles" just because of their juxtaposition in the aether field. These "illuminated" particles are both primary particles of matter and/or just fabric patterns, random, unstable patterns of aethers.

In other words, unless directly influenced by other particles, these temporary patterns don't have a unified vibration. In the cloud chamber, the aethers exist together because of their proximity and because of the impact of accelerated particles with set configurations.

To paraphrase how these phenomena have been recently described: out of the "vacuum energy" emerges particles which return into this vacuum. This vacuum energy could be single or two dimensional aether configurations. Is vacuum energy Dark Energy?

In the future, if, or when, a Unified Field Theory is established, cloud chambers will be a means to discover why certain elements exist and others are only configurations.

More In Depth Speculations

"Charge!"
(The Next Big Question)

"Charge," he shouted
Almost as a eureka moment.

"To where?" I replied,
"Against the enemy entropy?"

"No to the depths of the unknown,
To beyond where we can fathom
To the basic essence of it All
To the …
Charge."

"Why?"

"To find out what is:
All and nothing
Or
Positive and negative.
Charge!"

"Oh, to discover if it is
Mass-Energy and/or void,
Positive and/or negative."

"Yes, are these 4 entities?
Two entities?
Or one?
Charge!"

Exploring the Fundamental Substance

Without an essence of being
There would not be a thing

In one of my earlier papers concerning this theory, I hypothesized that the size of a single aether could be approximately 3.99×10^{-35} cm. But rethinking this, this is probably the size of a 3D mass because it is a Planck's Universal unit of least length derived from the equation $L^2p = hG/c^3$. Since a single aether is massless, pure energy then these units' sizes must be smaller and equal to their vibrations, which would be hG/c^3. These sizes will still vary depending upon their surroundings. But it may also be that Planck's constant, 6.626×10^{-34} joules per second, is the energy of a single, vibrating bipolar aether. Or if a single aether is "vacuum energy," then it could be smaller yet.

To say that the aether is positive and/or negative is too easy and really doesn't explain anything. But some fundamental substance definitely exists. Energy is undoubtedly that essence. Units of this energy are undoubtedly components of photons We know they exist, and they seem to be ubiquitous in the Universe.

Following are two possible explanations of their structure: A single aether can be a massless, single pure energy unit like Leibniz's monad. The positive aspect is the aether. The void, empty space, is the negative aspect.

Or what I am tending to believe these days, is that it is just positive and negative units: the Yin and Yang, the division which makes existence exist. This theory is implying that a negative energy force exists as a force in and of itself as does the positive. If these are so, then it is so.

Some Possible Positive and Negative Aether Structures

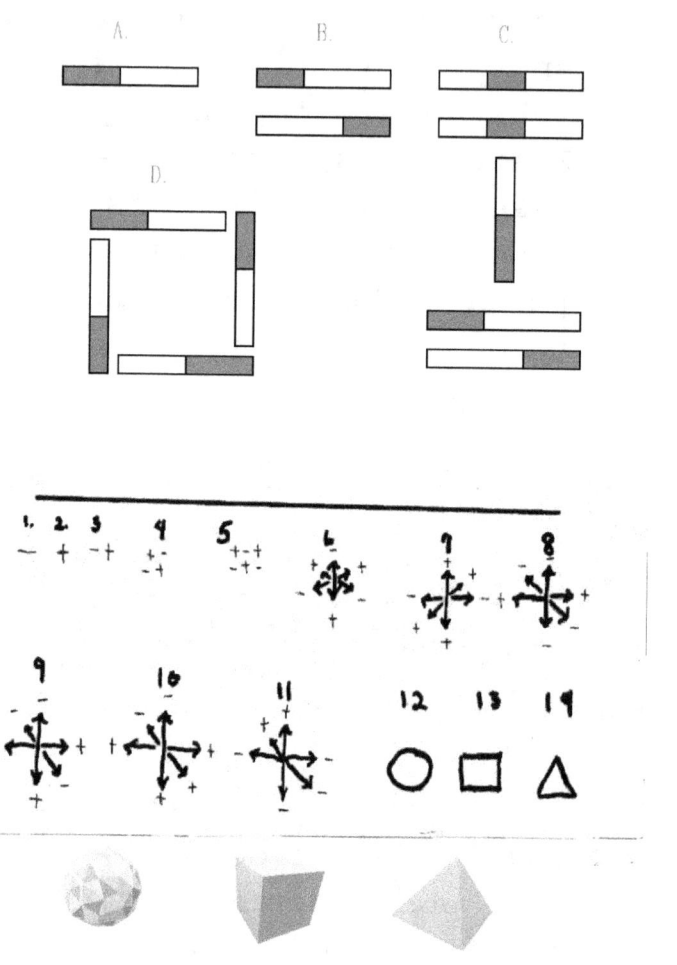

But if G. W. Leibniz is more correct and a type of monad exists, then the basic fundamental units of the Universe could be a vibrating unit of energy and a void, "a black hole." (Plank's least length?) The vibrating unit is the positive aspect and the void is the negative. This solves the problem of positive and negative energies colliding and the added dimension to prevent them.

So in this case, the Universe is energy and void. The vibration takes place because the energy wants to expand. Many units of energy try to fill a single void, and when a single void is filled with this energy, then the void of another black hole "attracts" the energy in that direction.

The vibrating energy creates a unit. The direction of vibration is indeterminate because the environment is random. It vibrates where it can.

In this scenario, mass and gravity take on a slightly different meaning. How can a void, a black hole, have mass?

In that aethers want to expand and black holes give them the capability, it appears as if the voids attract aethers.

Depictions of a Possible Black Hole Configurations:

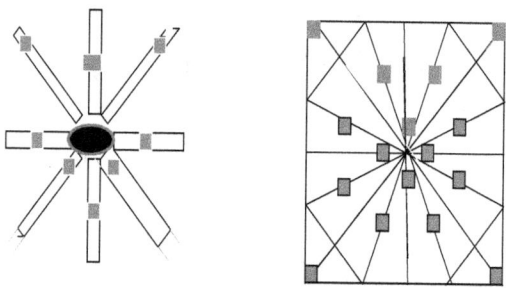

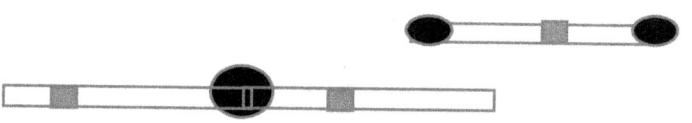

When aether configurations and structures surround these voids and can not penetrate them because they are all packed together, larger black holes are created. A structure of aether configurations create a shell surrounding the back hole

It's like too many people trying to exit through a too small a door; they get jammed. The following is a pie-wedge-view of a surrounded black hole:

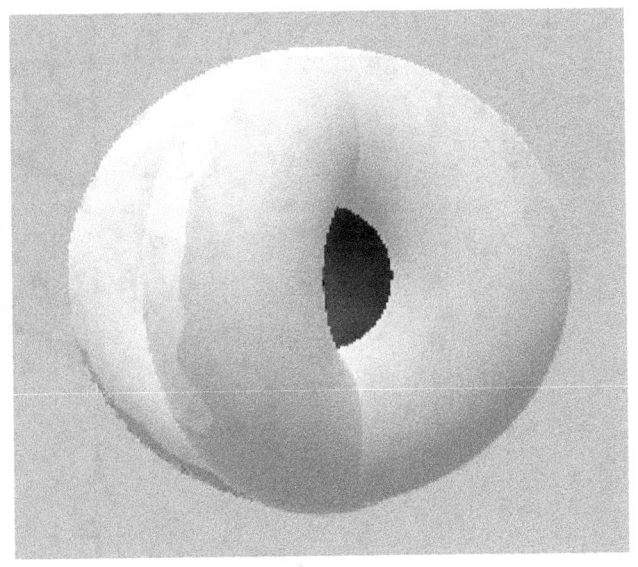

Now the main question: how is all matter created from such a situation? Electrons must have a structure where the energy units can not completely neutralize the voids. A void must be pushed forward by positive structures. See following:

Void   Aether   Bipolar

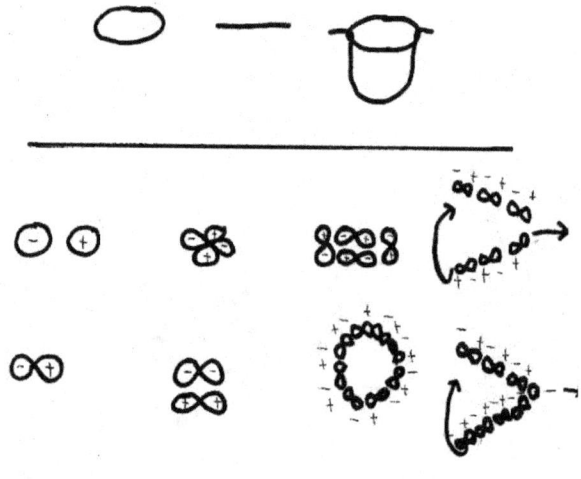

No matter which of these two depictions, positive-negative or energy-void, is more accurate, if either is, each of these primary units of aether is a single dimension when it vibrates, and only when combining with other units or traveling the speed of light do they add dimensions.

The third dimension, mass, emerges when two single dimensional objects (a combination of aethers on a $\underline{x}$ and $\underline{y}$ axes) are accelerated, or are combined and vibrating as on a *x, y,* and *z* axes. These configurations can explain why even assuming Super-symmetry, more positive particles exists in the Universe than negative. (see next)

How More Positive Particles Exist In Our Universe?

Preface

Zero to the zero power
Is equal to one
Therefore some thing exists?
Or
Some thing exists
Therefore zero to the zero power
Has to equal one?

How can anything exist?
If $0^0 = 1 = A$

And the charges of aethers are
$1 = +, -, \pm$

And $\pm A = +A$ or $-A$ or $\pm A =$ positive or negative or bipolar aethers

And the exponent function represents dimensions then the following could represent aethers in an aethered universe

An Aether	One dimensional aethers
1^1	1^2 = moving photons
-1^1	-1^2 = moving photons

Two aethers		Two dimensional aethers	
2^1	2^2 = 2D	2^3 = moving neutrinos	
-2^1	-2^2 = 2D	-2^3 = moving electrons	

Three aethers		Three dimensional aethers	
3^1	3^2	3^3	3^4 = moving particles
-3^1	-3^2	-3^3	-3^4 = moving particles

Explanation

If nothing to the zero power is **not** nothing then it must be something: a single dimension. (Nothing times nothing equals something? See comments concerning Math) The single dimensions are positive, negative, and bipolar aethers. Squaring these single dimensional units creates 2 dimensional entities, and if these are accelerated they become 3D. Cubing these single dimensional units creates 3 dimensional objects, and if these are accelerated they become 4D = length width and depth plus space-time, motion.

But when squaring the negative aethers they are no longer negative entities. If there is any truth in this facsimile, these equations, then it would explain why more positive particles exist in the universe.

Dark Energy and Dark Matter

Scientists are seeking an enormous mass in the universe which explains both why spinning galaxies rotate faster on their edges than in their middle, and the continued expansion of the universe. A aethered universe could go a long way to explain Dark Energy and Dark Matter.

Dark energy could be the single-dimensional, individual aethers: the positive, negative, and bipolar. Dark matter could be the two-dimensional aether structures. Single dimensional aether (pure energy) combines and creates two dimensional configurations (not mass and not not mass). We only recognize three dimensional configurations as particles, but in an aethered universe, one and two-dimensional structures would be far more plentiful than the three-dimensional. Especially if massless pure energy is the ubiquitous substance of this universe.

Using the present day accepted common knowledge that the science world is proposing: approximately 70% of the universe is Dark energy and approximately 25% is Dark matter with the remaining 5% being three-dimensional matter. There is approximately 14 times the number of single dimensional aethers and five times the number of two dimensional aether structures as there are three dimensional structures.

That all aethers and aether structures want to expand and yet gravity pulls everything back together explains the balance.

By calculating the amount of mass necessary to create this phenomenon, one may discover more exact properties concerning the aether.

Mathematics

Does or even can mathematics represent reality? Some say it is reality. To some it is another human creation, and nature only has mathematics incidentally involved with it.

Whether the Universe is infinite or not, math is. Many aspects of math are infinite and therefore can probably represent both all natural phenomena and all imaginary abstract thoughts.

I am amazed by people like Newton and Einstein who from what I can tell conceptualized reality then found formula to represent their concepts. Then coming from a different angle, it seems some people have been inspired to have understood the "truths" behind the representations of math and discovered a more encompassing reality.

In my mind, at most math can represent an ideal reality. Are we discovering reality, the one we can know, then looking for formula and equations to represent what we found, then manipulating numbers to discover more of what we can know? Are we confirming a possibility of our theories with math?

Through field theories one would probably explain and create mathematical equations for the aether, i.e. a variance of the gauge field equations.

Gravitons?

From the result of a catastrophic collision of two neutron stars, 130 million light-years away, gravitational waves were detected at 8:41 a.m. on Aug. 17. The gravity waves were followed in seconds by detection of high-energy gamma rays. Just 1.7 seconds after the gravitational wave signal, NASA's Fermi space telescope spotted a glimmer of gamma rays in the same neighborhood of the sky(Science

News Vol. 192 No. 8, November 11, 2017, p. 6) What phenomena caused the light to follow the gravitational wave?

Less Definitive Attempts at Proof or Disproof

Sagnac Phenomenon

By re-examining and re-interpreting some of the later Michelson's and Michelson - Morley type experiments (Post 475-478) they may indicate that indeed, the aether was detected. When a device is moved through a relatively stationary aether field, the time that the photons spend in this between-area should cause a variance in an angle of detection. In other words, any interferometer when it is in motion and the distance between the source of light and the reflecting surfaces is large enough so that it does not carry its own aura, then a resulting interference pattern should be created at the detector. (See MM Experiments) Or if the interferometer is accelerated and the aether field is stripped from the accelerated object(s) then fringe pattern shifts should be seen.

These are seen in the Sagnac phenomenon (Post 476) and when a ring laser and laser gyroscopes move through the aether.

The original experiments of the French physicist Georges Sagnac split a beam of light and sent the two separate beams in a different direction around a loop and recombined them. They should have recombined without an interference pattern because the aether around the earth is stationary; the whole apparatus rotates with the earth.

Then he rotated the whole platform on which the experiment rested. The interferometer did not register the change necessary to identify the aether because the results were consistent with special relativity

But in repeating the Sagnac's interferometer experiments of yesteryear what may be discovered is the initial readings when the instrument starts to rotate will have a greater fringe shift than a later reading. This may occur because the aether field within the interferometer might begin to spin with the instrument, although slippage will occur. My reasoning for this speculation is that when Pogany in his 1928 experiment placed two glass rods (which have their own stable aether field) "in the path of the light beam, (he) came within 1% of the theoretically expected fringe shift."(Post 477) The glass rods reduced the slippage of the aether field within the rotating interferometer.

Moon Laser Experiments

For years, scientists have been bouncing a laser light beam from reflectors on the moon.

By re-examining and re-analyzing these moon laser experiments of the 1960s and 1970s, one may find proof for the existence of the medium. This conclusion assumes that the magnetosphere is the polarized aether field and that the moon travels in and out of it.

The variables involved in the re-analysis are very complex: the rotating aether sphere of the earth, the magnetosheath "rushing passed" the earth, the rotating aether field of the sun, and the accelerating and de-accelerating moon traveling through these different aether domains. Without accounting for these different variables, a seemingly random collection of statistics were accumulated (Alley pp 368-370 & Silverberg p 219).

To determine, in retrospect, if these more random statistics were caused by the aether fields, one would have to calculate when the moon exited one domain for the next. Each transition from one field to the next should be indicated by a change in the angle of variance of the light to the moon and reflecting back, and what would appear to be an immediate fluctuation of the speed of the moon.

Since the moon travels in and out of the magnetosphere and the magnetosphere travels with the rotating earth, for the time that the laser light is in the magnetosphere, it travels with the earth both in the earth's direction around the sun and in the spin as the earth spins.

When the laser light leaves the magnetosphere, its forward momentum and rotation caused by the earth will cease and in turn will be affected by whatever field it enters.

If this theory has any validity, then the following phenomena should be observed:

1) As the moon enters the magnetosphere, any "lead[1]" (anticipation of the future path) for which the laser telescope compensated would cease and the moon would appear to slow. Also it follows that the closer the moon gets to magnetosphere it would need less and less of a "lead-time" because the laser light would be in the aether field for a larger proportionate time.

2) As the moon leaves the magnetosphere, the "lead" would have to be reapplied, and for a moment the moon would appear to accelerate. And

---

[1] The "lead" exists because the laser has to anticipate the mirror by the time it takes light to travel to the moon and the speed of both objects.

as the moon travels further away from the magnetosphere, the "lead" would need to increase because the laser light would be in the aether field a smaller proportionate time.

(This is not accounting for the possibility of the rotating aether field of the sun through which the earth is traveling or for "slippage" of the magnetosphere, the fact that these barriers are not exactly definitive.)

A smaller but still detectable phenomenon would be the "lead" needed to compensate for the rotating aether field of the earth.

All of these phenomena would be more pronounced if photons by themselves could be used, rather than laser beams. Laser beams don't ride the waves but plow the waves. Trying to measure an aether field with a laser beam is like trying to determine the direction of rain with an instrument placed on the nose of a fast-moving Bullet Train. (Possible but ………)

My conclusion about trying to determine the magnetosphere as the contained aether field has been delegated to "Less Definitive Experiments" because accurate results could possibly be obtained but too many factors are involved.

MMS - NASA's Magnetospheric Multiscale Experiments

As mentioned in the text, many past experiments, many re-observations, and many possible experiments could prove the existence of an aether medium. Any of various experiments where the light source and target are moving at relatively the same speed through a stationary aether should prove the existence of an aether: for the time which the photons are in the relatively stationary aether, they should not be moving at the same speed and in the same direction as the mechanism, and an angle of variance should be created. The source and the target will have to be far enough apart so that their surrounding aether fields do not overlap and far enough apart so that a deviance can be measured. This may be possible in deep space.

As I understand it, with the Magnetospheric Multiscale project and its experiments, measurements may be possible. As the satellites travel in and out of the Magnetosphere surrounding the earth, with constant monitoring of them, it may appear as if they accelerate or slow. While in the Magnetosphere where the aether is relatively stationary to the earth, especially at its trailing side, one speed should be registered. But when the satellites leave or reenter the magnetosphere, a different speed should be registered because of the speed of the spinning earth and its magnetosphere through the sun's magnetosphere.

If information is passed among the four satellites, if they are far enough apart, they will detect the aether rushing past them. These variances should change as they move in and out of the different aether fields, i.e. an adjustment of the instruments would be necessary to reconnect the sensors with each other.

(Lorenz) contraction
(This idea was created when I was still in Syracuse University 1972.)

All accelerated particles and objects should be pressured by the medium through which they move. As objects are accelerated through the aether field they should become "shortened" in the direction of flight. An example would be an object falling from space into the earth's gravitational fields. The closer it comes to the surface, the more it would be traveling through a more compressed aether field and thereby become more compressed. Falling to earth would compress the object by its length, and it would also form a "V" shape of aether field around it because of the gradually compressed aether field closer to earth.

A test for a contraction may be to spin two "Ts" (see following Figure as an example of one of the "Ts") such that the four electric sensors in the two cross bars are aligned and synchronized. Then change the rate of spin of one "T," such as stop it. If the second is still spinning then the electric sensors should not align, because the top of the "T" bar will have contracted. The sensors' length on the spinning "T" bar will have contracted. (Can these "T" bars be accelerated fast enough to test this?) If both "Ts" are spinning in opposite directions, and if the Lorentz Contraction exists both sets of sensors will align but the time when they encounter each other will (what?) take longer to align. As they begin to spin in opposite directions, they will align at a certain time, but then as they spin faster and faster, the timing of alignment will take longer because of the shrinkage and when the two bars are at maximum speed, they will again have a set time of alignment, but it will take longer(?) than when both were rotating in the same direction.

Figure

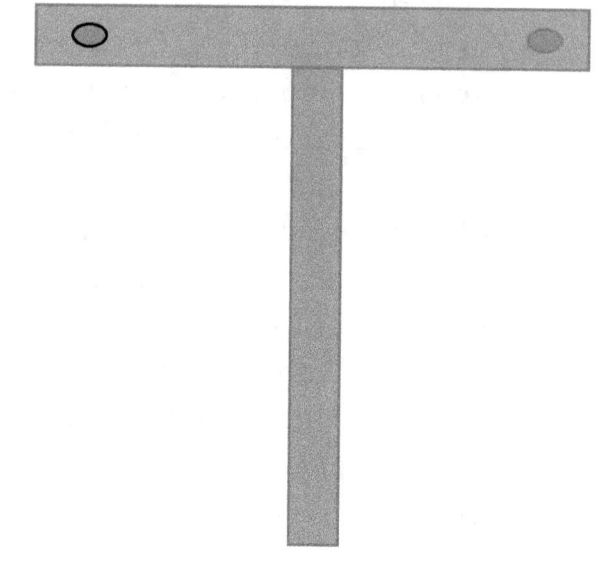

A Personal Experiment

On February 28th of 1998 in La Veta, CO I created an experiment as an "attempt to discover if a polarized aether exists" (this and the following quotes are from my MEAD Composition book titled <u>Magnetic - Photo flim(sp) Experiment.</u>)

The Question

To discover "If by spinning (polarized) aethers on their axes … it may be possible to create a photo(n)-like movement of the aethers," which can be detected.

This is assuming

"1) Polarized aether exists

2) A magnet can rotate it

3) That a magnet (mainly the magnetic which I had and its configuration) can rotate a photo(n)-like aether structure."

The Experiment

"A camera has been purchased but without a lens so that when the TMZp3200 film is inserted, the film is directly exposed to the environment. An arrangement has been created such that the shutter can remain open for an indefinite time."

A spinning magnet is placed at various distances from the exposed film to see if a dot (a light spot) is created by the spinning aether and is registered on the film.

"(4) Four 'rare earth' magnets were situated on a devise which could be placed on the end of the rotating rod from the electric motor. ...

"The shutter will be opened and the film exposed. Then the motor will be started by a push-on/push-off switch."

An example of a trial dated 3/6: "Exp(sure) 9....Opened shutter; turned on motor @ 2:20pm. @ 2:30 it was still running. @ 2:52 it was not running so clicked the shutter closed."

This series of trials lasted until 3/21 with approximately 36 exposures.

"2nd Attempt-

La Veta, Colorado
7/10/98 …

1) It was a question, a slight question, but still, if the film advanced on the first roll. It was also a question about the development of the film.
2) This time I'm going to try to move the 'polarized aether' by aligning the rotating magnets with the earth's poles. The mechanism will be point north then east on each attempt, @ 5 ½ inches … from rotating magnetic…."

This series of trials lasted until 7/12 with approximately 36 exposures.

"3rd Attempt

Am using Infrared Eir 135-36 film

could not change film in total darkness as requested on box so 1&2&3&4 –   ...

(Kept) in Icebox for a week. The temp in this room is 84° F."

This series of trials lasted until 7/16 with approximately 37 exposures.

The end result was no holes in the films, that I could see.

Conclusion

My first response to the lack of positive results was that I was wrong. No polarized aethers exist and I stopped thinking about it. But then, as findings from physicists and astronomers kept adding to and revising information, I kept returning to this theory because I thought I was right.

Also as I read more and more, I realized that I had no idea what I was doing with this experiment, and I had no idea what the aether was. My general assumptions could be correct, but the characteristics concerning the aethers could be wrong. I had no idea about the size of the individual aethers and therefore could not imagine how to influence it, or even if I had influenced it.

I still have all of the developed film from the experiments.

Conclusion to A Unified Field Theory

Forward

On the TV show *Bones*, Dr. Temperance"Bones"Brennan (Emily Deschanel) was addressing a group of scientists and she opened with a joke, which I have interpreted: Dr. Schrodinger was traveling to the beach when he was pulled over by a policeman. Upon seeing the doctor's driver's license, the state patrol said, "Oh yes, I've heard of you. You are the person with a cat in your trunk. May I open it to see?"

"Well I would prefer that you didn't…." but before the doctor could finish his sentence, the policeman opened it and announced, "You have a dead cat in your trunk," whereby everyone in the car yelled with a certain amount of frustration in their voices, "Now we do!!!"

So is there a problem in the question, moral or otherwise? Should the policeman not question and not open the trunk? Is it making a complicated Universe more complex? Do people who believe in multi-universes propose that if we don't open it we will be in a universe where we would never know, and that would be just fine?

No,… no folks once we know Dr. Schrodinger is driving and there is a rumor of a cat in the "boot" of his car, we as officers of the law, as scientists, as human beings, need to stop him and examine this trunk so we will know more about our Universe, about the "Now."

Even if we change the way the Universe is arranged by questioning, examining, opening different types of boxes, we need to open them, not just because we are curious, but because it is. And we know more about our Universe by doing it.

Why I Created This Theory and This Book

What is this manuscript? This is a human being's perspective, living at the turn-of-century in the United States of America. My educational background is a generalist, with a broad field of study. I am not a specialist in any single subject, save, maybe, art. I probably would not have created this theory if I would have been educated in the field of physics.

It may have been for the best for this theory because I was not indoctrinated into thinking a certain way, and I had no idea what the educational-system was teaching. When I took the Philosophy of Science class at Syracuse University and the results of the single and double slit experiments were described, I had no idea that electromagnetic waves could travel through empty space and that particles were appearing from nothing and could return to nothing in cloud chambers. I just thought - I can explain that. So I have tried.

By writing, I am recording my thoughts and ideas with the notion of furthering my perspective of human existence. I may actually be fulfilling a human function: therefore, the subtitle of the book. (We would probably have to ask an historian or even a psychologist concerning that.)

Yes, it is a human characteristic to try to understand the "whys" of the universe, but many human urges go unfulfilled and should. And as I have said, I am no physicist

So why I have been developing this theory and for forty years pursuing recognition of it is a question in itself (Especially when I have received little to no positive responses?) For one reason, I have really received no negative responses over the years from the physicists to whom I have presented it. Nobody has (bothered?) taken the time to say, "This is idiocy, and it is wrong because of this phenomenon. Or this experiment disproves your theory because……" And in many, many cases, no responses were forthcoming. Only occasionally a person would say, "Ok, quantify it."

The second reason … I really thought and continue to think that I have developed something: a Unified Field Theory. When I told my wife I that I was going to finally take the time and energy to sit down and write this manuscript because "I could possibly be right," she said she knew what I was going to say before I said it. She has probably heard that statement fifty times in the last twenty-five years because all of the new revelations in physics have encouraged me. I believe my theory is inadvertently in the process of being proven.

Over the last thirty or forty years, I have watched the interpretations of observed phenomena become closer to my Unified Field Theory: years ago all quanta functions took place in "empty space." But now, massless energy (vacuum energy?)

is evolving as a concept. Years ago, light and neutrinos were massless entities, which flew through empty space. Now the concept of particles emerging from the void and returning to the void can occur in Cloud Chambers. Now a Higgs field exists.

What are these massless entities from which particles can emerge? Units of energy that can come into existence and return to what? A void? No?

Even if this presented model of an aether is flawed as it probably is because of my lack of knowledge and misunderstandings of the results of experiments, concepts, etc. it should point the way to a Unified Field Theory, if such a thing exists.

Yes, it is true, very little of the material in the papers which have been submitted to many different sources have been quantitative. Personally, I am not qualified to quantify it nor even in a position to recheck past results from previous experiments, or to physically produce experiments to prove or disprove this theory, even though I have tried in all cases. The results of the experiments which are suggested in this paper only propose probable outcomes, and even though they are not precise figurative numbers, they are "gross" effects, which could still be measured by quantitative means.

I probably would not have spent the time and energy to get all of this material together if I had no outlet, but I can self-publish which is what I have had to do with all of my writings.

I believe that because of the Internet and our access its knowledge, our abilities to communication though it, and create (art objects?), it has empowered all of us. It has given power to the people, and this book is an example of what an individual is capable of doing. All I needed was the time and will and out came books and manuscripts which seem only I have wanted.

I know I am just writing for, pretty much, myself, generating my thinking, making clear my thoughts and ideas, precipitating my "knowledge," and organizing it all, to manifest manuscripts.

Yes, I am seventy-five years old, believing I am not getting any more intelligent, (hopefully wiser) and yet taking the Bull (lol) by the horns and pushing forward with this manuscript because all of my ideas and concepts have been scattered, and I wanted to get them together, and if I don't do it now, it will never get done.

Moreover, the worst-case scenario: I like my Science Fiction plausible and detailed.

The Final Justification

```
The Creator
  Created
  Creators
 To Create.
```

Epilogue

I would like to believe an epilogue exists for this manuscript: something will follow as a cause and effect. But I am not holding my breath.

Appendix

"The 'Hidden Variables' in the EPR Experiments."

(A Derivative of this paper was presented to *Physics Essays* in 1991. I would like to thank Dr. E. Panarella of the National Research Council in Ottawa, Canada for his reviews of this paper and early drafts. Even though he did not encourage this endeavor (understandably), his feed-back indicated that he understood my project, which was very important to me at the time because no one else even responded. He was very constructive and presented me with ideas and problems which the paper needed to confront.)

An Abstract

This paper proposes the possibility of the existence of an aether medium to explain data from the classical single and double slit experiments and from the A. Aspect et al experiments. Aspect, A.; J. Dalibard, and G. Roger, Phys. Rev. Letts 49, 1804 (1982) and Aspect, A.; P. Grangier, and G. Roger, Phys. Rev. Lett. 49, 91 (1982).

These latter experiments are an "optical version of the Einstein Podolsky Rosen Bohm Gedankenexperiment." (Aspect) This paper also proposes to re-examine the data produced and to propose future possible experiments.

Key words: aether medium, single and double slit experiments, Aspect experiments, quantum

An Explanation of Some Attributes of The Theory

The image I will create is designed to explain the EPR, Einstein, Podolsky, Rosen type experiments and the single and double slit experiments in an acceptable and understandable manner. Strange interpretations have been created to describe the results from these experiments. (See Original Rationalization).

Since this paper will try to explain quanta, I will use the example of an electron to represent matter, existing with mass and to represent virtual particles which can "appear and disappear," "the tunneling effect," when changing orbits.

Matter can move through the aether with little or no resistance because the forces holding any one aether in place are almost non-existent. That is to say in an idealized (deep) space, each positive and negative aether is surrounded on all sides by its opposite.

Electrons, in arranged, asymmetric aether fields around atoms move with little resistance because their orbits' configurations form to create a pattern of least resistance, as would any matter with long durational movement, which is to say the configuration of the aethers in front of a particle turn to create an attracting force.

Particles traveling the speed of light do not "pass through" the aether fields; only the "energy" of the particle does. For example, when an electron does its quantum jump from one orbit to the next, the electron as a particle does not move between orbits but disappears from one and reappears in the other: the energy and charge of the electron passes through the aether in the form of its configuration. An analogy here might be a tube filled with marbles. When a marble is shoved into one end, a different marble emerges from the other.

This is a displacement-compression wave. In the medium it moves instantaneously(?) or at the speed of light(?)… The orbits and fields of aether between the orbits are fairly stable. These intermediary fields are packed with aethers to the magnitude where they can not accommodate an electron. These fields' pressures are in part the "weak force" which keeps the nuclei from flying apart, and the electron from entering the nucleus, and the reason quanta are an all-or-none propositions.

Because of increased energy to the electron and the "strong force," the attraction of the protons in the nucleus, and momentary rifts occurring in the intermediate fields, the electron "collapses," and its configuration pounds into the field between orbits. These newly interjected aethers in the form of an electron have to displace aethers in the fields (the marble effect). The spinning two-dimensional configuration of the electron plows into the field, rearranging the field for a moment by displacing aethers, by spinning aether configurations, by interjecting a negative electrical charge which causes displacement and radiation to be emitted.

The rotating and rearranging of the intermediary fields create one dimensional photons. The momentary displacement-compression wave causes a compression wave to move inward then outward in all directions, like an expanding bubble. At the same time, the aether configurations which were set spinning [the photons] "ride" this wave.

A photon is a virtual particle with only its neutral spin traveling through the aether. When we "see" a photon particle, we are seeing the spiral- spinning of a relatively "stationary" aether configuration. It is like observing an extremely small spinning pinwheel, turning because of the wind. But in that one pinwheel turns, it causes another to turn, and so on. On a light sensitive plate, the photon becomes a scratched circle, a dot.

The spinning photon has a set parameter of freedom within the expanding bubble wave created by the compression and radiation. If another wave mingles momentarily and the amplitude builds or dampens, the parameters of motion [location] increases or decreases during that time. Also because no space is "idealized," the spinning configuration may not be totally compatible with the aether field directly in front of it, plus add to this the indeterminacy of photons travelling in waves because of their inability to penetrate other vibrating aethers, and the photon may vary slightly to one angle or another in any field.

Single and Double Slit Experiments

In 1802, when Thomas Young placed two pinholes in a piece of paper, he observed photons (the spinning configurations) plus the wave [compression and radiation] go through both holes, and the fragmented wave interfered with itself. When he closed one hole, the particles and waves went only through one hole: therefore, no interference pattern occurred. In later experiments, when only one photon and wave was released toward the double slits, the photon and wave went through one slit, and the wave went through the other, still creating an interference pattern. The same wave moving through two different slits skewed it slightly, causing it to interfere with itself.

In the single slit experiments, when the slit is approximately the size of the wavelength alternate light and dark bands, Fraunhofer diffraction patterns, plus dribble on both sides of the peak, are created. The wave's energy "will be absorbed and reradiated by the edges of the slit." (Besancon) This in part creates the dribble. But the aether pattern between the slit and the target also adds to the wavy dribble pattern. The photons travel in the aether's compatible-field pattern. In this case, the pattern could be a series of expanding half circles emerging from the single circle slit.

As the size of the slit, through which the light travels, is increased, the diffraction patterns are decreased because the number of photons which goes through the slit increases, thereby creating a more stable pattern in the aether field. Also increasing the intensity of the radiant energy, including using lasers, will change and stabilize this field pattern. This compatibility is important to understanding and predicting results from the

EPR Type Experiments.

When electrons are excited in the Aspect et al experiments Aspect, A.; J. Dalibard, and G. Roger, Phys. Rev. Letts 49, 1804 (1982) and Aspect, A.; P. Grangier, and G. Roger, Phys. Rev. Lett. 49, 91 (1982) and move to an outer orbit then collapse back into their previous orbits, twin photons can be created. When the twins are created, they expand outward in their single balloon-type compression radiation wave with correlating polarization. Even though these photons travel in a parameter of indeterminacy, within the wave, their polarizations are correlated because the original "pounding" in the aether field between the orbits creates not only the polarization but the expanding balloon wave with its polar configuration in which both travel, and at the same time, the balloon wave and photons influences the environment in front of the expanding wave. These in themselves are enough to create the correlations which the raw data provide.

In other words, the displacement-compression wave from the electrons changing orbits creates the photons and the compression radiation bubble wave. In that the bubble is an entity in itself, the aethers it encompasses as it passes through the field form a set configuration, and this in turn reforms the aether field in front of it. The pressure of the bubble and the photons riding in it "set up" the field in front.

The actual physical experiments were somewhat different than the EPR gedankenexperiment. Not just a single pair was released, but many pairs, a series of "cascades." So as mentioned earlier, just as the flowing photons in a single slit experiment could change the aether field, so could the flow of paired particles possibly change the pattern between the source and the target, thereby increasing the correlations. It may be possible to confirm this hypothesis by rechecking the results of the A. Aspect et al experiments.

From the beginning of a single cascade to the end, A) the number of hits should have increased proportionately to the number of projected paired particles, and B) the number of polarized correlations should have increased. As the experiment progressed A) and B) should also have been observed.

Again if this hypothesis is correct, because these photons exist in a parameter of indeterminacy and interference, as in the earlier ocean analogy, both particles will probably be less correlated the further away from each other they travel. For instance, as in the double slit experiment, obstacles can disintegrate the unity of the bubble: other waves and particles can interfere. But the bubble or balloon analogy breaks here because once the connection in the compression wave bubble disintegrates, the individual separate parts can still move outward from the source unless obstructed.

But undoubtedly while all the particles in the expanding balloon are connected in their moment of movement, it would be possible to change all of their polarizations by changing the polarization of one particle or one flow of particles.

If the axis of an aether structure or configuration is slightly shifted by a mechanistic means [traveling through a counter] and the parameter of indeterminacy is reduced, then all axes in the surface-circle would be altered. If this is true, the change in polarization would travel around the bubble (at the speed of light or instantaneously?) and yet the correlation between the twins would be maintained.

Even the Aspect (Aspect, A.; P. Grangier) experiment which switches polarized channels at 10ns would be affected. It may not be because the "signal" travels faster than the speed of light (although theoretically it could) but because the aether in the bubble and the aether field between the switches and target have been changed.

What I have been implying is that all these types of experiments could be reducing all the photons within a set parameter to a smaller parameter, and therefore the experiments, in fact, change the local aether patterns on the surface of the expanding bubble, which increases the correlations. Another factor involved, the aether field in front of the expanding bubble is altered by the configuration of the bubble and might change the counter before the bubble arrives because the wave field hits before the bubble.

A separate experiment could be created to test the above hypothesis. Rather than deploying an off-on switch, a polarized lens could be rotated while the photons are in motion. One, the experiment could start with two different lenses intercepting the two different streams of twin photons with the same polarization, and then one lens could be rotated. These results could be compared with the number of hits and correlations as the lenses re-align. While the slow rotating of the lens is occurring, not only should the number of hits on the target progressively decrease, but, probably, the correlations of polarization also. If the lens is stopped when both are re-aligned, the number of registered hits should increase because the turning filter will have "picked up" photons. In fact, a greater number of hits than in the Aspect et al. experiments could be recorded because the field between the source and target has been better formed.

As one lens is rotating, will the twin be rotating and polarization be maintained? If this hypothesis is correct then yes.

A separate experiment to test whether a polarized aether exists would be to produce a magnetic field between the slits and the target: a very strong field may change the pattern of the original field and a strong magnetic moving back and forth behind the target should create a more random pattern.

Bibliography of 'Hidden Variables'

Aspect, A.; J. Dalibard, and G. Roger, Phys. Rev. Letts 49, 1804 (1982).

Aspect, A.; P. Grangier, and G. Roger, Phys. Rev. Lett. 49, 91 (1982).

Einstein, A.; B. Podoisky, and N. Rosen, Phys. Rev. 47, 777 (1935).

Nussbaum, Allen in The Encyclopedia of Physics, 3rd Edition, Edited by Besancon, Robert M. (Van Nostrand Reinhold, N.Y, 1990) pp. 488-494.

Planck, Max. "The Theory of Heat Radiation" (pp 174-175): The History of Modern Physics. USA: Tomash Publishers, 1988 pp. 188-189.

Weidner, R. T. and R.L. Sells, Elementary Modern Physics (Allyn and Bacon, Boston, 1973) pp. 12-13.

The Original Rationalization

(Originally drafted in 1990)

Even though all of these concepts are suppositions, I maintain them because so many pointers seem to indicate their existence. I live in the physical world; I observe it and think about it. The explanations proposed by the world of physics with its gaping holes of nonexplanation, of randomness, of not being able to know, are almost mystical. It frightens me that we discuss our physical reality in such terms. It frightens me that we are experimenting with the subatomic particles, (the fabric of existence?) while we have just questionable explanations. Did anyone really know the results of exploding the Atomic bomb? It was detonated anyway.

I am no physicist, but neither were the early Greek atomists. Even if they were not completely correct in their assumptions, they pointed the way. Verification of atoms at that time was impossible, but yet, just through intuition they were close. It would be a gross miscalculation and extreme egoism to assume that humanity has progressed to a situation where only a step by step addition to present theories will contribute to our knowledge concerning the physical reality. This seems to be the same means by which the perfection of Ptolemy's universe was created. What an elaborate theory they had developed and how wrong. Even some of the early Greeks' intuitions were more accurate than all the facts, figures, and calculations of all those pre-Copernican scientists. Of course, Copernicus destroyed those rigid concepts of reality, but still the Ptolemaic universe had been preceded and superseded by Aristarchus of Samos and others.

It was Copernicus who saw the inconsistencies and the lack of explanations to phenomena which contradicted common sense, and so he must have said, "Enough is enough!" It seems to me that our reality as dictated by our present-day science is a Ptolemaic universe perpetrated by our system of education. We are so convinced that what science says is true, so convinced by the "facts" and the big names that all we can do is either stand in awe or add to this knowledge. Even the word "science" carries so much weight in this technological era that one would dare not lift it.

The metaphysical interpretations of the "raw data" of present day physics' experiments especially dealing with quantum mechanics [notable the Einstein, Podosky, Rosen types] range from an acceptance that "this is the way things are" to solipsism. The following are but a few of the interpretations of this "raw data": Particles, or even machines, "tell" other particles or machines how to respond without human intervention. Particles can know what other particles are doing. The speed of communications can exceed the speed of light. A poor cat can exist in limbo neither alive nor dead until we look at it. Observations can affect a reality which existed before the observation, and even our thoughts can reach into the past and change it. Even the more scientific interpretations which deal with indeterminacy, chaos, and multi-simultaneous, probabilistic universes if expanded and taken to their logical conclusions create universes in which it is hard to conceive we are living, or even want to live.

In my opinion, many of these interpretations seem to run counter to "common sense," "intuition," and even previously verified physical phenomena. The frightening aspect is that the more quantum mechanics seems to be understood, the more these types of interpretations pop out of the mass media, are accepted, and assumed to be true, not just the consumers of mass media but by scientists and philosophers, who understand the implications of the terms "theories" and "assumptions.

In my mind, this ontological hole has been deepening almost since the end of the nineteenth century. Developing this analogy, I would say this rift in reality is a mental black hole which is absorbing all the "facts," theories, questions, and interpretations generated but without emitting any real light. In fact, it seems that the more theories, interpretations, "facts," which are thrown into this abyss, the larger it grows because the unbelievable gains credence by consensus. Many scientists feel they have a grasp of this subatomic "reality" because they can predict the existence of particles and the outcomes of experiments. But how did the pre-Copernicans explain the "retrograde" motion of planets?

In that theory, the planets seem to stop, reverse direction, stop again without any visible cause and then resume their courses. Now we know that this explanation runs contrary to common sense, intuition, or any visible, known physical fact. Yet, it was accepted because scientists could predict the future positions of planets and other phenomena. Because of a retrospective perspective, we should understand that predicting results does not necessarily define reality.

Why in a universe of physical realities, which we can and do successfully manipulate, create, predict, and confirm results, could such anti-intuitive interpretations be generated and accepted? We are well enough attuned to the material world to have workable basic concepts. We as a species seem to be thriving. We are a creation of this Universe. But yet, the above-mentioned explanations and many other "modern" elucidations run counter to our pragmatic conceptions of this world

Our exotic interpretations, especially those of the mass media, originate in more than just our conceit. I feel we seek excitement, mystery, and even "magic" in our lives. And to satisfy these longings, all cultures, no matter how isolated or diverse, have created gods, and usually no simple god but the most complex, most powerful, most elaborate systems concerning these gods, and this impulse may indicate that mysterious labyrinths entangle humans not just to answer the unknown or for enjoyment but possibly to fulfill psychological needs. We seem to want to supply the gaps in our knowledge with stimulating speculations (this book for example? lol). Add to this that almost any interpretation will be tainted by our human biases, and the end results are off-shoots of interpretations which can be unnatural, mystical, and exotic beyond real.

Yes, our out of control imaginations (hooking us up with the infinite?) and psychological needs can help us to transcend our present perceptions, and even help us to know, reality, but as mentioned the Universe created us, we are part of it, and maybe we can search internally into our knowledge and intuition of existence to understand our existence and the universe.

To me it seems we have to revert back to winnowing our wheat, letting reality fall where it may and letting not only the chaff, but our fantasies and psychological needs be dispersed by the wind. Reality is truly weighty enough to withstand the windy weather, the new waves of exotic interpretations. Since we are subtly and inextricable connected to the metaphysical reality we seek, we should seek more plausible, more intuitive, more common-sense explanations using the known facts.

The existence of an aether should be re-thought and re-investigated. And maybe I should not have titled it "aether" because this term carries so much history. It connotes the classical medium by which light travels. But yet that is what it is, redefined. The contemporary scientists and/or philosophers might call it Vacuum Energy, but I feel it is what they were seeking around the turn of the century and what they are seeking now, so I will stick with the term "aether." By assuming this fundamental ubiquitous medium exists, a great deal of the ambiguity, inconsistencies, and unexplained phenomena which undermines physics today can be eliminated.

It is acknowledged that even if the accuracy of the specifies of this theory is lacking, by proposing that a medium exists, I hope to point the way to create a model that will demonstrate the plausibility of this type of theory. It is the simplicity and the vast number of phenomena that a theory like this can explain which makes the existence of a medium probable.

Referenced Works

Alley, C.O. et al. "Apollo 11 Laser Ranging Retro Reflector: Initial Measurements from the McDonald observatory." Science vol. 167, 1970, pp. 368-370.

Aspect, A.; J. Dalibard, and G. Roger, Phys. Rev. Letts 49, 1804 (1982).

Aspect, A.; P. Grangier, and G. Roger, Phys. Rev. Lett. 49, 91 (1982).

Bender, P. L.; Currie, D. G.; Dicke, R. H.; et al. (October 19, 1973). "The Lunar Laser Ranging Experiment" (PDF). Science 182 (4109): **229-**

Besancon, Robert M. The Encyclopedia of Physics, 3rd Ed. NY: Van Nostrand Reinhold, 1990.

Chapter 6. "Particle Spin and the Stern Gerlach Experiment" physics.mg.rdu.au

Chiao, R. Y.; P. G. Kwiat, and A. M. Steinberg "Faster than light?" *Scientific American*, August 1993).

Dicke, R.H. The Theoretical Significance of Experimental Relativity. NY: Gordon and Breach, 1968, p. 27.

Einstein, A.; B. Podoisky, and N. Rosen, Phys. Rev. 47, 777 (1935).

Hafele, J. C.; Keating, R. E. (July 14, 1972). "Around-the-World Atomic Clocks: Predicted Relativistic Time Gains." Science 177 (4044): 166-168.

Michelson, A.A. and E.W. Morley. Am J. Sci. 34, 333 (1887) 4.

Nussbaum, Allen in The Encyclopedia of Physics, 3rd Edition, Edited by Besancon, Robert M. (Van Nostrand Reinhold, N.Y, 1990) pp. 488-494.

Planck, Max. "The Theory of Heat Radiation" (pp 174-175): The History of Modern Physics. USA: Tomash Publishers, 1988 pp. 188-189.

Post, E.J. Reviews of Modern Physics. vol. 39 #2, pp. 475-478 (April 1969).

Science in 1972:
Hafele, J.C.; Keating, R. E. (July 14, 1972). "Around-the-World Atomic Clocks: Predicted Relativistic TimeGains". Science 177 (4044): 166-168.

Science News Vol. 192 No. 8, November 11, 2017, p. 6)

"Faster than light?" by R. Y. Chiao, P. G. Kwiat, and A. M. Steinberg in *Scientific American*, August 1993).

Silverberg, E.C. and D.G. Currie. "Lunar Laser Ranging with Decimeter Accuracy." Bull. Am. Astron. Soc. vol.4, num.2, Pt.1, 1972, p.219.

Weidner, R. T. and R.L. Sells, Elementary Modern Physics (Allyn and Bacon,Boston, 1973) pp. 12-13.

Other Works of William
In Chorological Order

Changing Planes and Shifting Gears

This novel examines the life of a nineteen-year-old boy who faces the Existential Dilemmas, contemplates suicide, but resolves the conflicts without reverting to religion. This book delves into how he found and created answers by which he could live and ultimately how he created meaning, purpose, and significance for himself.

A Fear and A Warning

Future possibilities in America are presented. It is the expression of a fear that organizations could become too involved in their citizens' lives, for power and control, so much so that individuals lose their guaranteed freedoms: freedoms that are innate to life itself and freedoms that are written into our social contracts. It is a warning to all human beings and societies not to misplace their values for self-gain or as an attempt to be "overzealous" for social causes.

Speculations about Being

This nonfiction manuscript is written by a person at the beginning of the twenty-first century who has tried to understand some of the major questions of this time in order to anchor a philosophy.

Under the Over the Counter

Experience and fiction merged to present a picture of the Over-the-Counter stock market and Penny Stocks at a certain time in our history. It is a study of characters.

An Odyssey of Speculations

As science fiction, it asks and speculates about some very basic questions and takes the reader to a possible future of our planet.

Search Books, Stories, and Songs

Eight manuscripts of poetry in chronological order.

www.ingramcontent.com/pod-product-compliance
Lightning Source LLC
Chambersburg PA
CBHW052159220526
45471CB00004B/1736